Water Science and Technology

CONTAMINANTS IN THE SUBSURFACE ENVIRONMENT

Organizing and Program Committee

P. L. McCarty *(Chair)*	USA
P. V. Roberts *(Co-Chair)*	USA
E. Arvin	Denmark
P. S. C. Rao	USA
R. Schwarzenbach	Switzerland
J. L. Wilson	USA
J. T. Wilson	USA
A. J. B. Zehnder	The Netherlands

Sponsors

IAWPRC
Western Region Hazardous Substance Research Center (Stanford University and Oregon State University)

Financial Supporters

US Environmental Protection Agency
US Geological Survey
US National Science Foundation

CONTAMINANTS IN THE SUBSURFACE ENVIRONMENT

Proceedings of the International Symposium on Processes Governing the Movement and Fate of Contaminants in the Subsurface Environment, held at Stanford University, California, USA, 23–26 July 1989

Editors

P. L. McCARTY and P. V. ROBERTS

Environmental Engineering and Science, Department of Civil Engineering, Stanford University, Stanford, CA 94305–4020, USA

PERGAMON PRESS

OXFORD · NEW YORK · BEIJING · FRANKFURT
SÃO PAULO · SYDNEY · TOKYO · TORONTO

U.K.	Pergamon Press plc, Headington Hill Hall, Oxford OX3 0BW, England
U.S.A.	Pergamon Press, Inc., Maxwell House, Fairview Park, Elmsford, New York 10523, U.S.A.
PEOPLE'S REPUBLIC OF CHINA	Pergamon Press, Room 4037, Qianmen Hotel, Beijing, People's Republic of China
FEDERAL REPUBLIC OF GERMANY	Pergamon Press GmbH, Hammerweg 6, D-6242 Kronberg, Federal Republic of Germany
BRAZIL	Pergamon Editora Ltda, Rua Eça de Queiros, 346, CEP 04011, Paraiso, São Paulo, Brazil
AUSTRALIA	Pergamon Press Australia Pty Ltd., P.O. Box 544, Potts Point, N.S.W. 2011, Australia
JAPAN	Pergamon Press, 5th Floor, Matsuoka Central Building, 1-7-1 Nishishinjuku, Shinjuku-ku, Tokyo 160, Japan
CANADA	Pergamon Press Canada Ltd., Suite No. 271, 253 College Street, Toronto, Ontario, Canada M5T 1R5

Copyright © 1990 International Association on Water Pollution Research and Control

All Rights Reserved. No part of this publication may be reproduced, stored in a retrieval system or transmitted in any form or by any means: electronic, electrostatic, magnetic tape, mechanical, photocopying, recording or otherwise, without permission in writing from the publishers.

First edition 1990

Library of Congress Catalog Card No. 82–645900

ISBN 0 08 040768 4

In order to make this volume available as economically and as rapidly as possible the author's typescript has been reproduced in its original form. This method unfortunately has its typographical limitations but it is hoped that they in no way distract the reader.

Printed in Great Britain by BPCC Wheatons Ltd, Exeter

CONTENTS

Preface	vii
P. S. C. RAO: Sorption of organic contaminants	1
J. A. PERLINGER, S. J. EISENREICH, P. D. CAPEL, P. W. CARR and J. H. PARK: Adsorption of a homologous series of alkylbenzenes to mineral oxides at low organic carbon content using headspace analysis	7
P. LAFRANCE, O. BANTON, P. G. C. CAMPBELL and J. P. VILLENEUVE: A complexation–adsorption model describing the influence of dissolved organic matter on the mobility of hydrophobic compounds in groundwater	15
P. M. BÜCHLER: The effect of exchangeable cations on the permeability of a bentonite to be used in a stabilization pond liner	23
J. DUCREUX, C. BOCARD, P. MUNTZER, O. RAZAKARISOA and L. ZILLIOX: Mobility of soluble and non-soluble hydrocarbons in contaminated aquifer	27
D. W. OSTENDORF: Long term fate and transport of immiscible aviation gasoline in the subsurface environment	37
J. B. CARBERRY and S. H. LEE: Fate and transport of petroleum in the unsaturated soil zone under biotic and abiotic conditions	45
J. M. THOMAS, V. R. GORDY, S. FIORENZA and C. H. WARD: Biodegradation of BTEX in subsurface materials contaminated with gasoline: Granger, Indiana	53
P. MORGAN and R. J. WATKINSON: Assessment of the potential for *in situ* biotreatment of hydrocarbon-contaminated soils	63
H. CLAUS and Z. FILIP: Enzymatic oxidation of some substituted phenols and aromatic amines, and the behaviour of some phenoloxidases in the presence of soil related adsorbents	69
K. MURAOKA and T. HIRATA: Basic study on TCEs behavior in subsurface environment	79
B. J. ALHAJJAR, G. V. SIMSIMAN and G. CHESTERS: Fate and transport of alachlor, metolachlor and atrazine in large columns	87
M. N. VISWANATHAN: Mineral sand mining and its effect on groundwater quality	95
J. GIBERT: Behavior of aquifers concerning contaminants: differential permeability and importance of the different purification processes	101
Subject index	109

PREFACE

The International Symposium on Processes Governing the Movement and Fate of Contaminants in the Subsurface Environment was held at Stanford University from July 23 to 26, 1989. The purpose of the symposium was to bring together leading scientists and engineers from around the world to evaluate current knowledge about the processes that affect the way in which chemicals interact with the soil, move with the flow of water and air through the subsurface environment, and are transformed or degraded by both abiotic and biotic processes. Basic understanding of such processes is necessary in order to develop realistic models for the movement and fate of contaminants in the environment.

At this meeting, there were presentations on eight special topics by invited speakers, twenty seven additional oral presentations, and forty four poster presentations. A total of 175 individuals attended the meeting, representing 15 different countries.

The symposium focused on the physical, chemical, and biological processes that are most important to chemical migration and transformation in the unsaturated zones above aquifers as well as in the groundwater zone. This includes information regarding both equilibrium and rate processes, and the various environmental variables that affect them. Among the processes of importance that were emphasized at the symposium were sorption, dissolution, volatilization, diffusion, multi-phase flow, and abiotic and biotic transformations. Experimental investigations were stressed, but presentations were also included that were oriented toward simulation of transport if they emphasized fundamentals of processes affecting the transport and fate of contaminants.

The preparation of a formal paper was not required by those making oral or poster presentations. The main objective was to gather together a knowledgeable group of individuals with interests in a common subject to discuss what is known and not known about the processes governing contaminant movement and fate. However, those wishing to present an original paper to be considered for publication were encouraged to do so. Their responses comprise this set of proceedings. In some ways it is unfortunate that the many other stimulating and useful thoughts that were exchanged at this meeting are not available in written form for sharing with those who did not attend the meeting. However, the organizers felt that it is important at times to also have meetings where writing of a paper is not required so that presenters are less constrained in presenting their ideas. We believe that this helped to attract an especially outstanding group of researchers. From the many comments received, we believe that the symposium was especially successful in meeting its objectives, and much valuable interchange of information occurred.

The symposium was jointly sponsored by the IAWPRC and the Western Region Hazardous Substance Research Center (Stanford University and Oregon State University). Partial financial support for the Symposium was provided by the U.S. Environmental Protection Agency, the U.S. Geological Survey, and the U.S. National Science Foundation. We appreciate the participation and support of these organizations.

Perry L. McCarty
Paul V. Roberts
Stanford University
Stanford, California, U.S.A.

SORPTION OF ORGANIC CONTAMINANTS

P. S. C. Rao

Soil Science Department, University of Florida, Gainesville, FL 32611-0151, USA

ABSTRACT

Sorption of organic contaminants plays a dominant role in their rate of transport through porous media such as soils, sediments and aquifers. The rate of abiotic and biotic transformations may also be significantly altered by sorption. This paper presents a brief overview of the current knowledge base on sorption of organic contaminants by natural sorbents, and examines several emerging issues that need further study. Partitioning into sorbent organic matter is viewed as the predominant process for sorption of nonpolar organics from aqueous solutions, and various methods have been proposed for estimating sorption/partition coefficients. Recent experimental and theoretical work has established a basis for predicting organic contaminants sorption from polar mixed solvents (mixtures of water and miscible cosolvents). Data on sorption of ionic and ionogenic organic compounds (e.g., phenols, amines) are limited, and initial efforts are underway to develop models for such sorbates. There also has been a recent resurgence of interest in studying vapor-phase sorption and transport of volatile organic compounds. Much of the early work on sorption dealt with equilibrium sorption, but considerable advances have been made in characterizing and predicting sorption nonequilibrium as well. Molecular-scale measurements are needed in order to provide direct, unequivocal evidence for selection among several contending phenomenological models proposed for describing sorption equilibrium and kinetics.

KEYWORDS

Groundwater contamination; vapor-phase sorption; complex mixtures; sorption kinetics; sorption nonequilibrium.

INTRODUCTION

The environmental behavior of organic contaminants in soils and groundwaters is governed by a number of coupled abiotic and biotic processes. These processes determine, either directly or indirectly, the rates at which contaminants applied at the ground surface migrate through the vadose zone and the amounts that are delivered subsequently to the saturated zone. These same processes, perhaps operating at different rates, determine the transport and attenuation of organic contaminants in the saturated zone as well. The fundamental difference between the two zones is that the saturated-zone pore spaces are completely filled with water (and at times certain nonaqueous phase liquids (NAPLs)), while the pore spaces in the vadose zone are occupied by water (or NAPLs) and air; these fluid phases each contain dissolved contaminants that can interact with the mineral and organic constituents of the solid phase. For the present discussion, the two primary processes of interest are solubility of organic contaminants in the liquid phase(s), and their subsequent sorption by the solid phases. Sorption is the primary process of retardation during contaminant transport through soils and aquifers, and also may influence the rates at which certain abiotic and biotic transformations occur. This paper presents a brief overview of the knowledge base describing sorption of organic contaminants by natural sorbents, and examines several emerging issues that need further study.

STATE-OF-THE-ART OF KNOWLEDGE

Over the past two decades, considerable progress has been made in understanding the various factors controlling sorption of organic contaminants by natural sorbents, and in developing conceptual models for predicting both sorption equilibrium and the kinetics of sorption. Among the numerous review articles and book chapters available on this topic, those of Karickhoff (1981, 1984) and Chiou (1989) provide the most comprehensive coverage of the various issues pertinent to equilibrium sorption.

Equilibrium Sorption from Aqueous Solutions

The term sorption has been used to denote uptake of organic solutes by a sorbent from a solvent, without reference to a specific mechanism. The prevailing hypothesis is that, for sorption of nonionic organic contaminants, organic matter is the dominant sorbent. Sorption is envisioned as a partitioning process, in which the sorbate permeates into the 3-dimensional network of the sorbent organic matter. In a majority of the cases, we are interested in sorption of predominantly nonpolar solutes from aqueous solutions. Thus, sorption may be viewed as the hydrophobic exclusion of a nonpolar (hydrophobic) solute and uptake by (or partitioning into) a hydrophobic sorbent. Chiou (1989), who is one of the strongest proponents of the partitioning hypothesis, summarized the supporting evidence as follows:

(1) linearity of sorption isotherms over a considerable concentration range;
(2) existence of an inverse, linear relationship between solute aqueous solubility (S_w) and sorption coefficient (K_p);
(3) generally low and exothermic heat of sorption (ca. 15-20 kJ/mole); and
(4) absence of competitive sorption when solutes are present in mixtures.

The partitioning model has been able to explain, at least in a phenomenological sense, a large amount of the data for sorption of a number of nonionic organic contaminants by a wide variety of natural sorbents (soils, sediments, and aquifer media). This model also serves as a conceptual basis for a number of equations proposed for estimating the carbon-normalized sorption coefficient, K_{oc}. Lyman (1982) has summarized and evaluated several of these regression equations. Hamaker and Thompson (1972), Rao and Davidson (1980), Kenega and Goring (1980), and Karickhoff (1981), among others, have compiled the published values of K_{oc} for a large number of organic compounds. The partitioning model is certainly not accepted universally without criticism (e.g., Mingelgrin and Gerstl, 1983). The limitations of this model for describing the sorption of organic contaminants with "reactive" functional groups and for sorption by low-carbon sorbents (e.g., aquifer solids) continue to be debated. The need to consider the concurrent role played by mineral constituents, and variations in the nature of organic carbon among different sorbents, has been argued by several investigators (e.g., McCarty et al., 1981; Reinhard and Curtis, 1984; Garbarini and Lion, 1986).

In contrast to nonpolar organic contaminants, current understanding of sorption of ionogenic and ionic organic solutes is fairly limited. Only recently has attention been focused on the sorption of organic bases and acids, such as substituted phenols, N-heterocyclic compounds, and other similar compounds. The degree of ionization of these solutes as influenced by sorbate pK_a and solvent pH, as well as the ionic concentration and composition of the solvent and the composition of sorbent, have been shown to have a profound impact on sorption of the ionogenic/ionic compounds (e.g., Zachara et al., 1987; Lee et al., 1989). Comprehensive summaries of the literature on this topic are, however, not available.

Sorption from Vapor Phase

In recent years, there also has been renewed interest in vapor-phase adsorption and advective-diffusive vapor transport of volatile organic chemicals (VOCs) in the vadose zone. The need to predict environmental behavior of VOCs at landfarms and at hazardous waste disposal/spill sites has been the primary motivation for the resurgence of research on vapor-phase behavior. Much of the early work on VOCs focused on pesticide volatilization from agricultural fields (see reviews by Taylor and Glotfelty, 1988; and by Glotfelty and Schomburg, 1989), but more recent studies have dealt with volatile constituents of gasoline and industrial solvents (Valsaraj and Thibodeaux, 1988; Poe et al., 1988; Rhue et al., 1988, 1989; Rao et al., 1989). Anhydrous clays and soils have a sizeable capacity to adsorb VOCs. Adsorption of volatile hydrocarbons, such as gasoline constituents, on "dry" sorbents is determined primarily by the available sorbent surface area, and the effects of specific sorbent-sorbate interactions appear to be minor in most cases. Water effectively outcompetes most of these VOCs, however, and therefore suppresses their adsorption. As a consequence, the adsorption coefficients for such VOCs decrease drastically with increasing relative humidity. Recent studies have shown that the heats of sorption for VOC sorption by anhydrous and moist sorbents are similar to their heats of vaporization (Rao et al., 1989).

Equilibrium Sorption from Complex Solvents

While sorption from aqueous solutions remains the major process of practical interest, considerable insight has been gained by examining sorption from mixed solvents containing mixtures of water and water-miscible organic solvents. Furthermore, it has been argued that, at or near waste-disposal or spill sites, the solution phase is more likely to be a mixture of water and several organic solvents. The effects of cosolvents, termed *cosolvency*, on sorption can be predicted by the log-linear cosolvency model proposed by Rao et al. (1985). The fundamental basis for this model is the observed log-linear increase in the solubility of nonionic organic solutes as the volume-fraction of the cosolvent is increased; this, in turn, leads to a decrease in sorption. This model has been experimentally verified by a number of recent investigators (e.g., Nkedi-Kizza et al., 1985, 1987), lending further support to the argument that sorption of nonionic organic solutes is primarily an entropy-driven process, and that solute activity in the solvent is the dominant factor determining sorption by natural sorbents.

Effects of Colloids

In most studies dealing with the environmental behavior of organic contaminants in soils and groundwater, the presence of colloids in the liquid phase has not been explicitly considered. One class of colloids are inorganic and organic particles with diameters less than 10 μm. Colloids found in soils and groundwaters include: dissolved organic carbon (DOC), such as humic acids and surfactants; biocolloids such as microorganisms; microemulsions of NAPLs; and mineral precipitates, such as oxides and sulfides of metal cations. The colloidal materials have similar sorptive properties as do the solid matrices of soils and aquifer solids from which they originate. Inorganic and organic contaminants become bound to such colloids, and may be transported in the aqueous phase along with the colloids. Thus, the mobility of inorganic and organic contaminants that are otherwise thought to be immobile can be enhanced considerably when colloids are present in significant quantities. McCarthy and Zachara (1989) have summarized available literature on the genesis, stabilization, and transport of colloids in groundwater, and their role in enhancing contaminant transport.

Sorption Nonequilibrium

The assumption of local equilibrium does not always provide an adequate description of organic contaminant sorption under dynamic flow conditions in porous media. The local-equilibrium assumption has come under increasing scrutiny in recent years. Brusseau and Rao (1989a), Pignatello (1989) and Harmon et al. (1989) have each presented thorough reviews of the experimental evidence for the various factors controlling sorption nonequilibrium, and of several modelling approaches that have been proposed to date.

The processes responsible for nonequilibrium may be grouped into two general classes: (1) transport-related; and (2) sorption-related. The first arises when sorption at the sorbent-solvent interface is itself "instantaneous" but access to the sorptive domains is rate-limited. Advective solute transport is essentially negligible in certain domains/regions of some porous media. The rate of mass transfer between these "immobile" domains and those in which advective-dispersive solute transport dominates is limited by solute diffusion within pore sequences of the "immobile" domains. This type of nonequilibrium is important in soils that are either aggregated or macroporous, as well as in fissured/fractured aquifers and in heterogeneous aquifers with layers and inclusions. A second type of nonequilibrium results because of rate-limited solute interactions with specific sorption "sites", including solute diffusion within the polymeric matrix of sorbent organic matter. The former is likely when sorption is the result of site-specific interactions (e.g., chemisorption), while recent evidence suggests that the latter is likely to be operative for "hydrophobic" sorption where the sorbate partitions into organic matter.

While a number of empirical approaches for estimating equilibrium sorption coefficients (K_p, ml/g) are available, the ability to estimate sorption rate constants (k, hr^{-1}) which describe nonequilibrium sorption has been limited. Based on an analysis of literature data, Brusseau and Rao (1989b) noted that k and K_p were inversely related, and that this dependence for nonionic organics can be described by the relationship:

$$\log k = 0.301 - 0.668 \log K_p; \quad n=61, \quad r^2=0.95 \tag{1}$$

The inverse k-K_p relationship for sorption of organic contaminants with "reactive" functional groups could also be described by a log-log linear function with similar slope, but with a much smaller intercept:

$$\log k = -1.789 - 0.62 \log K_p; \quad n=12, \quad r^2=0.40 \tag{2}$$

The above equations suggest that, the more hydrophobic is the solute, the slower will be its rate of sorption. Note that, at a given K_p, the sorption rate constant, k, for polar or ionogenic compounds is almost 30-fold smaller than the k value for nonpolar organic solutes. Brusseau and Rao (1989b) proposed that the inverse dependence is a consequence of intra-organic matter diffusion of the sorbate being the rate-limiting step. The diffusion of sorbates with "reactive" functional groups is further constrained by interactions with the functional groups of the organic matter; thus k is smaller. More recent studies have provided further support for the above equations and have tested their validity in providing excellent estimates of sorption nonequilibrium in lab-scale column studies (Brusseau et al., 1989a). Pignatello (1989) also concluded that diffusion is the main rate-limiting step for sorption, although he suggested that intra-particle diffusion may also be an important factor. Recent experimental studies by Ball (1989) also support this argument.

Another factor contributing to sorption nonequilibrium is aquifer heterogeneity, as manifested by the presence of discrete layers or inclusions of differing hydraulic conductivity and/or sorptive capacities. Transport-related nonequilibrium arising from such heterogeneities is likely to be a more dominant factor at field-scale than is intra-organic matter mass transfer (Brusseau and Rao, 1989c; Brusseau et al., 1989b).

FUTURE RESEARCH NEEDS

Additional data on equilibrium and nonequilibrium sorption of ionogenic and ionic organic contaminants need to be collected. These contaminants are found in a variety of energy-related wastes (e.g., coal and petroleum wastes), but a paucity of data has limited our ability to develop conceptual models for the sorption of these compounds. Predicting sorption of such contaminants is likely to require much more complex models than for nonionic "hydrophobic" organic compounds. While many data have been collected for simple systems involving sorption of single sorbates from aqueous solutions, much remains to be learned about sorption from complex systems involving mixtures of multiple solutes and multiple solvents. The significance of colloids, dissolved organic carbon and suspended particulate matter, in mediating the transport of organic solutes has just been recognized; further research on this topic is essential if we are to understand the long-range transport of strongly-sorbed solutes (e.g., PCBs, TCDD, etc.). Even though various conceptual models are available for predicting nonequilibrium sorption, our ability to estimate the necessary model parameters has been limited. Progress made in recent years needs to be continued. Furthermore, inferences regarding the processes controlling sorption equilibrium are based on large amounts of data from lab-scale experiments, but molecular-scale observations providing *direct* evidence are essentially nonexistent. This situation needs to be remedied by increased focus on spectroscopic and other relevant methods for studying sorption on natural sorbents. The relative significance of various processes and factors contributing to sorption nonequilibrium at varying spatial and temporal scales needs to be better understood.

ACKNOWLEDGEMENT

Approved for publication as Florida Agricultural Experiment Station Journal Series No. ____.
This study was supported, in part, by funds from Florida Department of Environmental Regulation contract WM-254.

REFERENCES

Ball, W.P. (1989). Equilibrium sorption and diffusion rate studies with synthetic organic chemicals and sandy aquifer materials. Ph.D. Dissertation, Stanford Univ., Stanford, CA.

Brusseau, M.L. and Rao, P.S.C. (1989a). Sorption nonideality during organic contaminant transport in porous media. *CRC Crit. Rev. Env. Control*, 19, 33.

Brusseau, M.L. and Rao, P.S.C. (1989b). Evidence for diffusional mass transfer within sorbent organic matter as a cause for sorption nonequilibrium. *Chemosphere*, 18, 1691.

Brusseau, M.L. and Rao, P.S.C. (1989c). Nonequilibrium and dispersion during transport of contaminants in groundwater: Field-scale processes. In: *Contaminant Transport in Groundwater*, H.E. Kobus and W. Kinzelbach (Eds), A. A. Balkema, Rotterdam, Netherlands, pp 237-244.

Brusseau, M.L., Jessup, R.E. and Rao, P.S.C. (1989a). Modeling the transport of solutes influenced by multi-process nonequilibrium. *Water Resour. Res.*, 25(10), (in press).

Brusseau, M.L., Sudicky, E.A. and Rao, P.S.C. (1989b). Solute transport under nonideal conditions. (Abstract), *EOS, Trans. Amer. Geophys. Union*, 70, 342.

Chiou, C.T. (1989). Theoretical considerations of the partition uptake of nonionic organic compounds by soil organic matter. In: *Reaction and Movement of Organic Chemicals in Soils*, B.L. Sawhney and K. Brown (Eds). SSSA Special Publication No. 22, Soil Sci. Soc. Amer., Madison, WI, pp 1-30.

Garbarini, D.R. and Lion, L.W. (1986). Influence of the nature of soil organics on the sorption of toluene and trichloroethylene. *Env. Sci. Technol.*, 20, 1263.

Glotfelty, D.E. and Schomburg, C.J. (1989). Volatilization of pesticides from soil. In: *Reactions and Movement of Organic Chemicals in Soils*, B.L. Sawhney and K. Brown (Eds). SSSA Special Publication No. 22, Soil Sci. Soc. Amer., Madison, WI, pp 181-207.

Hamaker, J.W. and Thompson, J.M. (1972). Adsorption. In: *Organic Chemicals in the Environment*, C.A.I. Goring and J.W. Hamaker (Eds), Mercel Dekker Inc., NY, pp 51-139.

Harmon, T.C., Ball, W.P. and Roberts, P.V. (1989). Nonequilibrium transport of organic chemicals in groundwater. In: *Reaction and Movement of Organic Chemicals in Soils*, B.L. Sawhney and K. Brown (Eds). SSSA Special Publication No. 22, Soil Sci. Soc. Amer., Madison, WI, pp 405-438.

Karickhoff, S.W. (1981). Semi-empirical estimation of sorption of hydrophobic pollutants on soils and sediments. *Chemosphere*, 10, 833.

Karickhoff, S.W. (1984). Organic pollutant sorption in aquatic systems. *J. Hydraulic Eng.*, 110, 707.

Kenega, E.A. and Goring, C.A.I. (1980). Relationship between water solubility, soil sorption, octanol-water partitioning, and concentration of chemicals in biota. In: *Aquatic Toxicology*, J.G.A. Eaton, P.R. Parrish, and A.C. Hendricks (Eds), Special Techn. Pub. No. 707, Amer. Soc. Testing & Materials, Philadelphia, PA, pp 78-115.

Lee, L.S., Rao, P.S.C., Nkedi-Kizza, P. and Delfino, J.J. (1989). Influence of solvent and sorbent characteristics on distribution of pentachlorophenol in octanol-water and soil-water systems. *Env. Sci. Technol.*, (in review).

Lyman, W.J. (1982). Adsorption coefficients for soils and sediments. In: *Handbook of Chemical Property Estimation Methods*, W.J. Lyman et al. (Eds). McGraw-Hill Book Co., NY, pp 4.1-4.32.

Mackay, D.M., Ball, W.P. and Durant, M.G. (1986). Variabilities of aquifer sorption properties in a field experiment on groundwater transport of organic solutes: Methods and preliminary results. *J. Contam. Hydrol.*, 1, 119.

McCarthy, J.F. and Zachara, J.M. (1989). Subsurface transport of contaminants. *Env. Sci. Technol.*, 23, 496.

McCarty, P.L., Reinhard, M. and Rittman, B.E. (1981). Trace organics in groundwater. *Env. Sci. Technol.*, 15, 40.

Mingelgrin, U. and Z. Gerstl. (1983). Reevaluation of partitioning as a mechanism of nonionic chemicals adsorption in soils. *J. Environ. Qual.*, 12, 1.

Nkedi-Kizza, P., Rao, P.S.C. and Hornsby, A.G. (1985). Influence of organic cosolvents on sorption of hydrophobic organic chemicals by soils. *Env. Sci. Technol.*, 19, 975.

Nkedi-Kizza, P., Rao, P.S.C. and Hornsby, A.G. (1987). Influence of organic cosolvents on leaching of hydrophobic organic chemicals through soils. *Env. Sci. Technol.*, 21, 1107.

Pignatello, J.J. (1989). Sorption dynamics of organic compounds in soils and sediments. In: *Reaction and Movement of Organic Chemicals in Soils*, B.L. Sawhney and K. Brown (Eds). SSSA Special Publication No. 22, Soil Sci. Soc. Amer., Madison, WI, pp 45-80.

Poe, S.H., Valsaraj, K.T., Thibodeaux, L.J. and Springer, C. (1988). Equilibrium vapor phase adsorption of volatile organic chemicals on dry soils. *J. Hazardous Materials*, 19, 17.

Rao, P.S.C. and Davidson, J.M. (1980). Estimation of pesticide retention and transformation parameters required in nonpoint source pollution models. In: *Environmental Impact of Nonpoint Source Pollution*, M.R. Overcash and J.M. Davidson (Eds), Ann Arbor Sci. Publ., Inc., Ann Arbor, MI, pp 23-67.

Rao, P.S.C., Hornsby, A.G., Kilcrease, D.P. and Nkedi-Kizza, P. (1985). Sorption and transport of hydrophobic organic chemicals in aqueous and mixed solvent systems: model development and preliminary evaluation. *J. Env. Qual.*, 14, 376.

Rao, P.S.C., Ogwada, R.A. and Rhue, R.D. (1989). Adsorption of volatile organic compounds on anhydrous and hydrated sorbents: Equilibrium adsorption and energetics. *Chemosphere*, 18, 2177.

Reinhard, M. and Curtis, G.P. (1984). A two-phase model for sorption of hydrophobic organic solutes by organic matter and inorganic surfaces. Paper No. 01F17, Abstracts of the Int. Chem. Congress of Pacifiic Basin Societies, Honolulu, HI, Dec. 1984.

Rhue, R.D., Rao, P.S.C. and Smith, R.E. (1988). Vapor-phase adsorption of alkylbenzenes and water on soils and clays. *Chemosphere*, 17, 727.

Rhue, R.D., Pennell, K.D., Rao, P.S.C. and Reve, W.H. (1989). Competitive adsorption of alkylbenzene and water vapors on predominantly mineral surfaces. *Chemosphere*, 18, 1971.

Taylor, A.W. and Glotfelty, D.E. (1988). Evaporation from soils and crops. In: *Environmental Chemistry of Herbicides*, R. Grover (Ed), CRC Press, Inc., Boca Raton, FL, pp 89-129.

Valsaraj, K.T. and Thibodeaux, L.J. (1988). Equilibrium adsorption of chemical vapors on surface soils, landfills, and landforms: A review. *J. Hazardous Materials, 19,* 79.

Zachara, J.M., Ainsworth, C.C., Cowan, C.E. and Thomas, B.L. (1987). Sorption of binary mixtures of aromatic nitrogen heterocyclic compounds on subsurface materials. *Env. Sci. Technol., 21,* 397.

ADSORPTION OF A HOMOLOGOUS SERIES OF ALKYLBENZENES TO MINERAL OXIDES AT LOW ORGANIC CARBON CONTENT USING HEADSPACE ANALYSIS

Judith A. Perlinger*, Steven J. Eisenreich*, Paul D. Capel**, Peter W. Carr*** and Jung Hag Park†

*Department of Civil and Mineral Engineering, Environmental Engineering Sciences; ***Chemistry Department, University of Minnesota, Minneapolis, MN 55455, USA
**United States Geological Survey, Water Resources Division, St. Paul, MN 55101, USA
†Department of Chemistry, YeungNam University, Gyongsan, 713-749 Korea

ABSTRACT

A homologous series of alkylbenzenes was sorbed to well-characterized mineral oxides in batch experiments to elucidate mechanisms of sorption in the absence of natural organic matter. Sorbed concentration was determined by measuring the headspace concentration in a closed vessel partially filled with a water/alkylbenzene/solid slurry. Partitioning of dissolved alkylbenzenes to sorbed alkylbenzenes was observed at high sorbed concentrations. At low sorbed concentrations, the mechanism was surface adsorption limited by the surface area of the solids. Enhancement of the adsorption of individual compounds was observed when alkylbenzenes were sorbed as a mixture.

KEYWORDS

Sorption; hydrophobic organic chemicals; mineral oxides; headspace analysis;

INTRODUCTION

The extent to which hydrophobic organic contaminants will sorb to solids is expressed in most mathematical models of contaminant transport in terms of a partition coefficient, $K = C_s/C_w$, where C_s is the equilibrium sorbed concentration and C_w is the equilibrium aqueous concentration. Measurements of sorption coefficients have often been made in water/solid systems with appreciable amounts of natural organic matter (e.g. sediment or soil slurries; Hamaker & Thompson, 1972; Karickhoff, 1981). For these systems, the entropy change, and thus the free energy change, ΔG_{cav}, resulting from the collapse of water's cavity structure when the organic molecules are transferred into the solid's natural organic matter, is thought to drive the sorption reaction (Rao et al., 1985). This partitioning of the chemicals between the aqueous solution and the organic matter on the solids has been empirically modelled as a function of the chemical's hydrophobicity and the amount of natural organic matter present:

$$\log K = a \cdot \log K_{ow} + b \cdot \log f_{oc} + c$$

where K_{ow} is the chemical's octanol-water partition coefficient and f_{oc} is the fractional organic carbon content of the solids (e.g., Schwarzenbach & Westall, 1981; Karickhoff, 1984). The free energy of the sorption reaction, ΔG_{ads}, is actually composed of four terms:

$$\Delta G_{ads} = \Delta G_{cav} + \Delta G_{elect} + \Delta G_{vdw} + \Delta G_{hb}$$

where ΔG_{cav} is as defined above, ΔG_{elect} is the energy resulting from electrostatic interaction between the organic molecule and the solid surface (dipole-dipole); ΔG_{vdw} is the energy resulting from association of molecules with each other and with the surface (induced dipole-induced dipole or dipole-induced dipole); and ΔG_{hb} is due to hydrogen bonding between the hydrated surface and the molecule. In systems where the amount of organic carbon is low, specific interactions between the chemical and the mineral surface contribute to the sorption reaction, and the last three free-energy terms take on greater importance. The relative importance of these specific interactions is not well understood (Karickhoff, 1984; Curtis et al., 1986; Rao et al., 1985). Aquifers, in general, have dissolved organic carbon concentrations of 2 mg/L or less (Thurman, 1985) and fractional organic carbon contents of < 0.001. Characteristics of the solids that may influence the sorption coefficient in this case are particle surface area, mineral composition (e.g. SiO_2, Fe_2O_3), structure (e.g. porous particles, expandable clay), and surface charge. Characteristics of the chemical that may influence the sorption coefficient are chemical hydrophobicity, polarizability, and size (e.g. total surface area, molar volume). Solution properties (e.g. pH, ionic strength, temperature) also may influence sorption reactions by affecting surface characteristics of the solids as well as the activity of the chemicals. This study seeks to understand the mechanism of adsorption of hydrophobic organic contaminants in systems where organic carbon content is low.

MATERIALS AND METHOD

In general, the sorption experiments involve equilibrating a mixture of alkylbenzenes (Table 1) in aqueous solutions containing well characterized mineral oxide particles. The particles used in the experiments reported here were bentonite (a smectite clay) and corundum (α-Al_2O_3). They have mean diameters of 6 μm and specific surface areas of 31 and 0.55 m^2/g, respectively, as determined by $N_{2(g)}$ BET sorption isotherms. Particles were combusted at 500° C overnight to oxidize any trace organic matter.

Table 1. ALKYLBENZENE PROPERTIES.

R	Vapor press (25°C)* (atm)	Solubility (25°C)* (g/m^3)	Recom. K_H (25°C)* (*10^3 atm*m^3/mol)	log K_{ow}
–H	0.125	1780	5.4±0.25	2.13
–CH_3	0.0375	515	6.6±0.35	2.65
–CH_2CH_3	0.0125	152	7.9±0.7	3.13
–$(CH_2)_2CH_3$	4.43×10^{-3}	55	6.9±3.0	3.69
–$(CH_2)_3CH_3$	1.32×10^{-3}	12.6	13.0±2.5	4.28

*Mackay & Shiu, 1981.

In the headspace analysis method, the fugacities of chemicals in all phases (gas, liquid, solid) within a closed vessel are determined by measuring the headspace concentration and using Henry's law and mass balance relations. This method is similar to the EPICS method (Lincoff & Gossett, 1984), except that only one vessel is used. For a closed vessel partially filled with a water/solid slurry, the mass balance for a chemical added to the vessel and allowed to equilibrate between the three phases is:

$$M_{total} = M_{gas} + M_{liquid} + M_{sorbed}$$

$$M_{total} = C_g V_g + C_l V_l + C_s M_{solids}$$

where C_g = gas concentration; V_g = volume of gas in the vessel; C_l = aqueous concentration; V_l = liquid volume; C_s = sorbed concentration (mass contaminant/mass solids); M_{solids} = mass of solids in the vessel. Using Henry's law constant $K_H = C_g/\tau C_l$, the mass balance can be rewritten as:

$$M_{total} = C_g V_g + C_g V_l/\tau K_H + C_s M_{solids}$$

Because C_g is measured and M_{total}, M_{solids}, V_w, and V_g are known, the equation can be solved for C_s:

$$C_s = (M_{total} - C_g V_w/\tau K_H - C_g V_g) M_{solids}^{-1}$$

The aqueous activity coefficient (τ) was determined to be unity at the ionic strengths and aqueous concentrations of the experiments reported here (Perlinger, 1989). The headspace analysis method eliminates phase separation and compound extraction steps necessary in batch sorption experiments, thus increasing the precision and rapidity of the analysis procedure. Accuracy is also increased with headspace analysis because incomplete phase separation, which can occur as a result of centrifugation or filtration (Gschwend & Wu, 1985, Morel & Gschwend, 1987), is eliminated.

A headspace analyzer was designed after Hussam & Carr (1985) to be used for the sorption experiments (Figure 1). The major components of the analyzer are a thermostatted, continuously stirred, glass reaction cell (volume = 4 L, 98% filled with slurry), a capillary GC/FID for sample analysis, and a gas transfer system. All gas transfer lines are 1/16" O.D., 0.01" I.D. precleaned nickel and are heated to 200° C. The gas sample is exposed to glass, stainless steel, and nickel in this system. The alkylbenzenes are injected into the cell with a syringe through a stopcock on top of the cell. The amount injected is determined gravimetrically. A 12.5 L glass bulb containing a known mass of compound provides a gaseous standard for analyses.

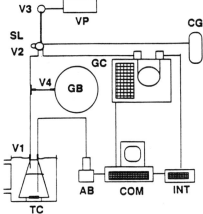

Figure 1. Headspace analyzer. SL = sample loop; V1,V2,V4, = air actuated valves; V3 = toggle valve; VP = vacuum pump; GB = gas bulb; GC = gas chromatograph; CG = carrier gas; TC = thermostatted cell; AB = autoburette; COM = computer; INT = integrator.

VERIFICATION OF METHOD

To verify that the aqueous activities of chemicals could be accurately and precisely measured by this method, air concentrations were measured after equilibrating additions of an alkylbenzene mixture in a known volume of water in the cell. The mass of solute dissolved in water was computed as the total mass minus the mass in the air. The slope of the regression line of the partial pressure (proportional to the measured air concentration according to the ideal gas law) vs. water concentration is equal to the Henry's law constant (Figure 2). Be-

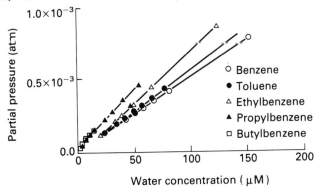

Figure 2. Titration of an alkylbenzene mixture into the cell partially filled with water. The slope of the line for each compound is the measured Henry's law constant.

cause the slope is constant over the entire concentration range and has a value that is within the range of values reported in the literature (Perlinger, 1989), and because the intercept of the line is zero, it is clear that the alkylbenzenes are not participating in reactions other than volatilization when only air and water phases are present in the cell. Reactions such as sorption to the glass container or association of the alkylbenzenes with one another do not occur. This conclusion is important because it indicates that when alkylbenzenes are added to a suspension of particles in the cell, only two reactions occur: volatilization and sorption. If other reactions were occurring, additional terms would have to be added to the equations above, and the system would be more complex. Values of Henry's law constants were identical regardless of whether individual compounds or a mixture of alkylbenzenes were added to the cell (Perlinger, 1989). This again confirms that the alkylbenzenes do not co-associate in the water when added as a mixture to the cell at these concentrations. Relative standard deviations ≤ 3% are achieved for the Henry's law constants.

SORPTION EXPERIMENTS

The first sorption experiment involved varying the bentonite suspended solids concentration from 0.016 mg/L to 9.75 mg/L while keeping the total amount of an equimolar mixture of alkylbenzenes, each initially 3.3×10^{-5} mole/L, constant in the cell (Figure 3). Samples were taken after 6 hours equilibration time, since earlier experiments had shown that equilibrium was obtained after 200 minutes (Capel, 1986).

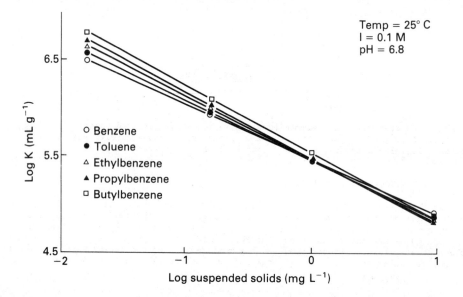

Figure 3. Log K vs. log Suspended solids concentration [SS]. The reversal in the relative magnitude of log K for the 5 alkylbenzenes is significant and may result from a steric hindrance of the larger alkylbenzenes to sorption.

Sorption coefficients decreased with increasing suspended solids concentration, as has been observed previously in laboratory studies (O'Connor & Connolly, 1980). The most hydrophobic compound, butylbenzene, had the highest sorption coefficient at low suspended solids concentrations. At higher concentrations, the order reversed such that benzene was sorbed most strongly. The observed decrease in the sorption coefficient with increasing suspended solids concentration was surprising in that analytical artifacts that can cause the decrease were absent. The solids were not physically separated from the water, so inaccuracies caused by incomplete phase separation (Morel & Gschwend, 1987) were not present.

The reason for the decrease in the sorption coefficient with increasing suspended solids can be better understood by plotting the sorption coefficient vs. fractional organic carbon content of the particles and surface coverage (Figure 4). The fractional organic carbon (g C/g

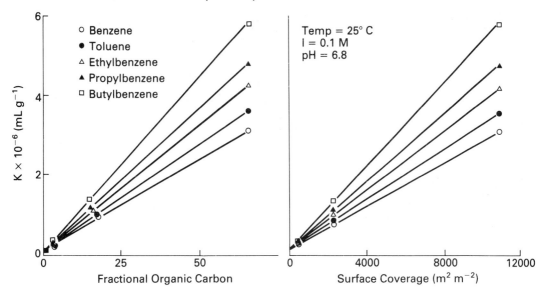

Figure 4. The influence of f_{oc} and surface coverage on the sorption coefficient. Sorbed alkylbenzenes cause high f_{oc} values and they form multilayers on the particles. The mechanism of sorption appears to be partitioning of dissolved alkylbenzenes to sorbed alkylbenzenes.

solid) that results from sorbing alkylbenzenes to the mineral surface can be computed as:

$$f_{oc} = C_s \cdot N \cdot MW$$

where C_s is the concentration of sorbed alkylbenzene (mole/g), N is the moles of C atoms per mole of alkylbenzene, and MW is the molecular weight of carbon. When a mixture of alkylbenzenes is sorbed the total f_{oc} is simply the sum of the contributions of f_{oc} of each of the alkylbenzenes. Similarly, the surface coverage of alkylbenzenes sorbed to the mineral surface can be computed as:

$$\text{surface coverage} = C_s \cdot TSA/(2 \cdot AREA)$$

where C_s is the concentration of sorbed alkylbenzene (mole/g), TSA is the total surface area of the alkylbenzene (m^2/mole), and AREA is the specific surface area of the solids (m^2/g). The total surface coverage resulting from sorbing a mixture of alkylbenzenes is again equal to the sum of the individual contributions of surface coverage. It is assumed in this computation that the alkylbenzenes sorb parallel to the mineral surface. A surface coverage of one indicates monolayer coverage. Total surface areas of the alkylbenzenes were computed by R. Dickhut (College of William and Mary) according to the method of Pearlman (1980). The sorption coefficient increased linearly with f_{oc} and surface coverage (Figure 4). At the same f_{oc} or surface coverage value, the sorption coefficient increased with compound hydrophobicity. All f_{oc} values were greater than one, and the surface coverage values indicated multilayer coverage. We conclude that the sorption reaction occurring under these conditions is one of partitioning of the alkylbenzenes in solution to alkylbenzenes already sorbed to the mineral surface. When sorption coefficients were predicted from octanol-water partition coefficients and f_{oc} (Schwarzenbach & Westall, 1981), the measured sorption coefficients were 1.5 to 2.5 orders of magnitude greater than those predicted. Alkylbenzenes in solution appear to partition more readily to sorbed alkylbenzenes than to natural organic matter.

To study interactions of alkylbenzenes with mineral surfaces rather than with sorbed alkylbenzenes, further experiments were carried out with higher suspended solids concentrations and lower aqueous concentrations. The resultant surface coverages were much lower (<1). In this experiment benzene alone was added to the cell containing water having 2.75 g/L corundum

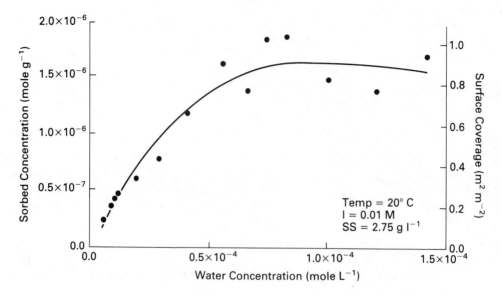

Figure 5. Benzene/corundum sorption isotherm. The sorption isotherm has a Langmuir shape and levels off at surface coverages of one, indicating that the available mineral surface limits the extent of adsorption.

concentration. Under these conditions (Figure 5), the sorbed concentration initially increased linearly with aqueous concentration, had a zero intercept, and leveled off at a surface coverage of about one, thus exhibiting a Langmuir-type adsorption isotherm. The leveling off of the isotherm at surface coverages of one indicates that the amount of surface available limited the extent of the sorption reaction. A linear regression of the points in the initial portion of the isotherm yielded a sorption coefficient of 24 mL/g (n=7, r^2=0.976).

To determine if the presence of other alkylbenzenes would influence the adsorption of benzene even at surface coverages < 1, an equimolar mixture of alkylbenzenes was added to the cell instead of benzene alone (Figure 6). Sorption isotherms of the individual compounds again exhibited Langmuir isotherms, but surface coverages leveled off at values less than one. Total surface coverage leveled off at a value of one, however, again indicating that the extent of adsorption was limited by the amount of surface available. The most hydrophobic compound (butylbenzene) had the lowest sorption coefficient, as was observed at the highest suspended solid concentrations in Figure 3. One possible explanation is that the larger (more hydrophobic) molecules are more sterically hindered from sorbing than the smaller (less hydrophobic) molecules when sorbed as a mixture from aqueous solution. A sorption coefficient of 94 mL/g (n=8, r^2=0.936) was computed from the linear regression of the initial portion of the benzene isotherm. The higher sorption coefficient for benzene in this experiment indicates that the presence of other alkylbenzenes enhanced the sorption of benzene, even at surface coverages less than one. This implies that the use of individually-determined sorption coefficients to model sorption of solutes from an aqueous solution containing a mixture of compounds is problematic because the solutes influence one another, at least at surface coverages > 0.1.

The value of the sorption coefficient for benzene sorbing to corundum (24 mL/g) is higher than previously reported values for hydrophobic organic chemicals sorbing to low f_{oc} solids, (typically < 1 mL/g). The hypothesis currently being investigated is that the surface coverages in these experiments (≥ 0.1) are significantly higher than those typically used (10^{-5} to 10^{-2}) in batch or column sorption experiments with low f_{oc} solids. The fact that the presence of other chemicals enhanced the sorption of benzene at surface coverages between 0.1 and 1 suggests that self-enhancement also may occur. This would result in increasing sorption coefficients with increasing surface coverage. It is important to understand the sorption reaction over the full range of surface coverages, because a wide range in surface coverages results when organic contaminants leach into the saturated zone, dissolve, and are transported with the ground water.

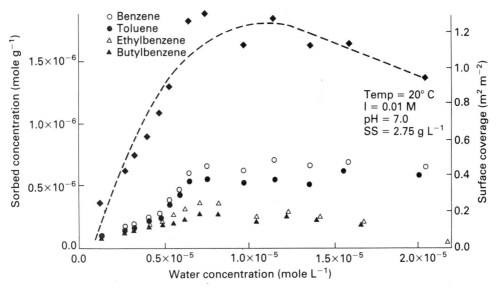

Figure 6. Alkylbenzene mixture/corundum sorption isotherm. The individual sorption isotherms level off at surface coverages less than one. The total surface coverage (filled diamonds) levels off at surface coverages of one, however, indicating again that the available mineral surface limits the extent of adsorption.

SUMMARY

The behavior of alkylbenzenes sorbing to mineral oxides observed in these studies is summarized in a hypothetical sorption isotherm (Figure 7). At high aqueous concentrations, many

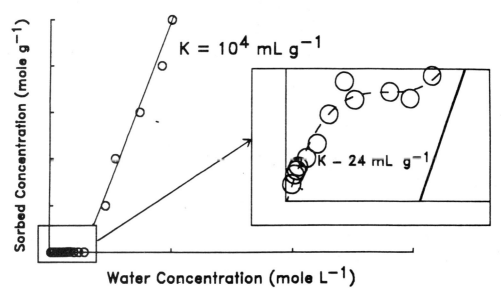

Figure 7. Hypothetical alkylbenzene/mineral oxide sorption isotherm. Partitioning appears to be the sorption mechanism at high sorbed concentrations (> monolayer coverage), whereas surface adsorption appears to be the mechanism at low sorbed concentrations (< monolayer coverage).

layers of alkylbenzenes sorb and the sorption reaction is one of partitioning of the alkylbenzenes in solution to sorbed alkylbenzenes on the surface. The resulting sorption coefficients are even higher than would be computed by an equation predicting the sorption coefficient when natural organic matter is present in equivalent amounts on the mineral surface. At lower aqueous concentrations (expanded in the inset in Figure 7), Langmuir-type isotherms are observed, and sorbed concentrations level off at a surface coverage of one. Because the amount of available surface limits the extent of the reaction, the mechanism must be surface adsorption in this concentration range. The presence of a mixture of alkylbenzenes enhances the adsorption of benzene, however, indicating that even at surface coverages less than one there is an influence of other organics on the surface adsorption reaction.

REFERENCES

Capel, P.D. (1986). Distributions and diagenesis of chlorinated hydrocarbons in sediments. Ph.D. thesis, University of Minnesota.

Curtis, G.P., Reinhard M., and Roberts P.V. (1986). Sorption of hydrophobic organic compounds by sediments. In: Geochemical Processes at Mineral Surfaces, Advances in Chemistry Series, American Chemical Society.

Gschwend, P.M., and Wu, S.C. (1985). On the constancy of sediment-water partition coefficients of hydrophobic organic pollutants. Environ. Sci. Technol, 19, 90-96.

Hamaker, J.W., and Thompson, J.M. (1972). Adsorption. In: Organic Chemicals in the Environment, vol. 1, , C.A.I. Goring and J.W. Hamaker (Eds.) Marcel Dekker, Inc., New York, NY, 49-143.

Hussam, A., and Carr, P.W. (1985). A study of rapid and precise methodology for the measurement of vapor liquid equilibria by headspace gas chromatography. Anal. Chem., 57, 793-801.

Karickhoff, S.W. (1981). Semi-empirical estimation of sorption of hydrophobic pollutants on natural sediments and soils, Chemosphere, 10, 833-846.

Karickhoff, S. (1984). Organic pollutant sorption in aquatic systems. J. Hydraulic Eng., 110, 707-735.

Lincoff, A.H., and Gossett, J.M. (1984). The determination of Henry's constants for volatile organics by equilibrium partitioning in closed systems. In: Gas Transfer at Water Surfaces, W. Brutsaert and G.H. Jirka (Eds.), D. Reidel Publishing Co., Dordrecht, Holland, 17-25.

Mackay, D., and Shiu, W.Y. (1981). A critical review of Henry's law constants for chemicals of environmental interest. J. Phys. Chem. Ref. Data, 10, 1175-1199.

Morel, F.M.M., and Gschwend, P.M. (1987). The role of colloids in the partitioning of solutes in natural waters. In: Aquatic Surface Chemistry, W. Stumm (Ed.), Wiley-Interscience, New York, NY, 405-422.

O'Connor, D.J., and Connolly, J.P. (1980). The effect of concentration of adsorbing solids on the partition coefficient. Water Res., 14, 1517-1523.

Pearlman, R.S. (1980). Molecular surface areas and volumes and their use in stucture/activity relationships. In: Physical Chemical Properties of Drugs, S.H. Yalkowsky, A.A. Sikula, and S.S. Valvani (Eds.) Marcel Dekker, Inc., New York, NY, 331 pp.

Perlinger, J.A. (1989). Sorption of a homologous series of alkylbenzenes to mineral oxides, M.S. thesis, University of Minnesota.

Rao, P.S.C, Hornsby, A.G., Kilcrease, D.P., and Nkedi-Kizza P. (1985). Sorption and transport of hydrophobic organic chemicals in aqueous and mixed solvent systems: model development and preliminary evaluation, J. Environ. Qual., 14, 376-383.

Schwarzenbach, R.P., and Westall, J. (1981). Transport of nonpolar organic compounds from surface water to groundwater. Laboratory sorption studies, Environ. Sci Technol., 15, 1360-1367.

Thurman, E.M. (1985). Organic Geochemistry of Natural Waters, Martinus Nijhoff/Dr. W. Junk (publ.), Dordrecht, Holland, 14-15.

A COMPLEXATION–ADSORPTION MODEL DESCRIBING THE INFLUENCE OF DISSOLVED ORGANIC MATTER ON THE MOBILITY OF HYDROPHOBIC COMPOUNDS IN GROUNDWATER

P. Lafrance, O. Banton, P. G. C. Campbell and J. P. Villeneuve

INRS–Eau, Université du Québec, 2800 rue Einstein, Suite 105, Québec, Qué., G1X 4N8, Canada

ABSTRACT

Natural dissolved organic matter found in groundwater can bind hydrophobic contaminants to form "complexes" and possibly affect their transport in the subsurface. The mobility of trace organic contaminants in soil originating from non-point source pollution may thus be affected by complexation reactions in the unsaturated zone. In order to predict the possible impact of such interactions on contaminant retention in soil, simulations have been made with a three-site sorption model. Two kinetic rate equations and an equilibrium Freundlich equation are used to describe possible sorption of two species of the contaminant (free or bound with dissolved organic matter) in a soil-water system. The equations governing the contaminant adsorption and the transport are simultaneously solved using an explicit-implicit finite difference technique, under steady-state water flow conditions. An analysis of the model's sensitivity to variations in the "complex" sorption rate constants and the complexation constant, K_c, demonstrated the relative importance of these processes and their effects on the vertical movement of the contaminant in soil. The influence of slow sorption kinetics for the "complex" varies as a function of residence time in the soil column, i.e. the pore-water velocity. Sorption of the "complex", when it does occur, diminishes the "carrier-effect" of the complexation. If the "complex" is non-adsorbable on soil, the transport of contaminants with K_c values greater than 10^5 (mol/L)$^{-1}$ will be significantly affected by dissolved organic matter concentrations typically encountered in soil-water systems. The possible application of the model to chemical transport studies in soil is discussed with respect to our present knowledge of complexation processes that can occur in field situations.

KEYWORDS

Groundwater; contamination; hydrophobic compounds; adsorption; complexation; dissolved humic substances; transport model; sensitivity analysis.

INTRODUCTION

The mobility of hydrophobic organic contaminants (HOC) in the unsaturated zone of soils depends mainly on their partitionning between the aqueous phase and natural sorbents, namely the immobile soil matrix and the mobile colloids. It is well recognized that sorption of nonpolar synthetic compounds is closely related to the organic carbon content of the soil or suspended particles. Many recent laboratory investigations have shown that dissolved organic matter (DOM), such as naturally encountered in interstitial soil water, may also play an important role in the distribution and the migration of HOC in the environment. The extent of association or binding of HOC by natural DOM such as humic or fulvic acids has been quantified for some polycyclic aromatic hydrocarbons or PAHs (McCarthy and Jimenez, 1985; Gauthier et al., 1986), polychlorinated biphenyls or PCBs (Hassett and Milicic, 1985; Caron and Suffet, 1987; Keoleian and Curl, 1987) and pesticides (Carter and Suffet, 1983). Such effects, as well as studies on the nature and concentration of DOM found in groundwater (Thurman et al., 1982; Paxéus et al., 1985; Thurman, 1985), raise the possibility that DOM may affect the transport of HOC in soils (Thurman and Malcolm, 1983).

In parallel with such experimental studies, environmental concerns for groundwater contamination have prompted the use of deterministic models to represent processes responsible for HOC transport towards the water table. Despite the increasing use of unsaturated zone transport models to predict the fate of HOC, relatively few studies (Bengtsson et al., 1987; Enfield and Bengtsson, 1988) have proposed to quantify possible effects of DOM on the hydrodynamic transport of HOC in the subsurface. In this study, we present a conceptual model for describing HOC movement in soil (Lafrance et al., 1987; 1989); the model assumes instantaneous reversible complexation by DOM in the aqueous phase of a fraction of the contaminant. The model further assumes that two species of the contaminant (free and bound to DOM) are adsorbed on three groups of sites in the soil. Simulations were carried out in order to evaluate the potential effects of the extent of complexation and sorption kinetics on the breakthrough curves (BTCs) for HOC through a water saturated soil column, under steady-state water flow conditions. This analysis of the model's sensitivity to variations in key physico-chemical parameters delimits the theoretical conditions for which a DOM-assisted transport for HOC could affect the contamination of groundwater.

THEORETICAL DEVELOPMENT

The proposed model (Lafrance et al., 1987; 1989) describes a non-competitive sorption on soil of two forms of a pollutant in the presence of DOM, e.g. humic substances (H). Two groups of sorption sites are assumed to be available for the free form (P) - one group achieves instantaneous equilibrium (Sp1), whereas adsorption on the second group is slower (Sp2). A third group of sites is assumed to be available for the bound form of the pollutant (HP) and is also controlled by slow kinetics (Shp).

Assuming that complexation in the aqueous phase is reversible and instantaneous, and defining K_c (mol/L)$^{-1}$ as the equilibrium constant for the complexation, one may write:

$$H + P \rightleftharpoons HP \qquad K_c = (HP) / [(H)(P)] \qquad (1)$$

where quantities in parentheses are the aqueous concentrations (mol/L) of the reactants and the product of the reaction. Given mass conservation of the total quantity of pollutant: $(P_t) = (P) + (HP)$, the fraction of the free form, α_p, and of the bound form, α_{hp}, of the pollutant may be written:

$$\alpha_p = (P) / (P_t) = K_c^{-1} / [K_c^{-1} + (H)] \qquad (2)$$

$$\alpha_{hp} = (HP) / (P_t) = K_c(H) / [K_c(H) + 1] \qquad (3)$$

From relationships between quantities of pollutant sorbed by soil, Sp1, Sp2 and Shp, and pollutant concentrations in solution, P and HP (Lafrance et al., 1987; 1989), the rate of change in the total adsorbed phase concentrations with time (k_i being the sorption rate constants) may be written:

$$\frac{\partial S}{\partial t} = \frac{\partial Sp1}{\partial t} + \frac{\partial Sp2}{\partial t} + \frac{\partial Shp}{\partial t} \qquad (4)$$

$$\frac{\rho}{\theta} \frac{\partial S}{\partial t} = \left[\frac{\rho}{\theta} KN1(P)^{N1-1} \right] \frac{\partial(P)}{\partial t} + \left[k_1(P) - \frac{\rho}{\theta} k_2 Sp2 \right] + \left[k_3(HP) - \frac{\rho}{\theta} k_4 Shp \right] \qquad (5)$$

The vertical movement of HOC subject to adsorption (but not degradation) under steady-state soil-water conditions is assumed to obey the following differential equation:

$$R \frac{\partial C}{\partial t} + \frac{\rho}{\theta} \frac{\partial Sp2}{\partial t} + \frac{\rho}{\theta} \frac{\partial Shp}{\partial t} = D \frac{\partial^2 C}{\partial x^2} - v \frac{\partial C}{\partial x} \qquad (6)$$

where R is the retardation factor, C is the concentration of the solute in units of mass per volume of solution (M/L^3), S is the adsorbed phase concentration in units of mass per mass of medium (M/M), D is the dispersion coefficient (L^2/T), v is the average pore-water velocity (L/T), ρ is the bulk density of the medium (M/L^3), θ is the soil-water content (L^3/L^3), x is the linear distance (L) and t is the time (T).

The differential equations were solved using numerical solutions with an explicit-implicit finite difference approximation method (Selim et al., 1976).

SENSITIVITY ANALYSIS OF THE MODEL

Conditions for simulations

The data for Equation (6), used to simulate BTCs for displacement of the pollutant through a soil column, are given in Table 1. The injection was maintained for the first half of the simulation time.

TABLE 1   Data for the soil column used in the simulations

Column length, z (m):	1.0
Average pore-water velocity, v (m/h):	0.02
Dispersivity, A (m):	0.05
Bulk density, ρ (g/cm^3):	1.5
Soil-water content, θ (cm^3/cm^3):	0.4
Input pollutant concentration (mol/L):	10^{-7}
Dissolved humic substances concentration (mol/L):	10^{-5}
Distribution coefficient, K (cm^3/g):	variable
Constant of complexation, K_c (mol/L)$^{-1}$:	variable
Rate constants, $k_1 ... k_4$ (1/h):	variable

Simulations were performed using a concentration value of the contaminant of 10^{-7} mol/L and a concentration of dissolved humic substances of 10^{-5} mol/L. The first condition corresponds, for example, to a pesticide concentration (with a typical molecular weight of 250) of 25 µg/L in the interstitial water of soil. The second condition represents a realistic concentration of dissolved organic carbon (DOC) in the unsaturated zone of 10 mg/L (Thurman, 1985). Such a value is obtained with the favourable assumption that all the DOC in groundwater can be described as humic matter with an average molecular weight of 2000 (Thurman et al., 1982) and a carbon content of 55% (Thurman and Malcolm, 1983). The value of $K = 20$ corresponds, for example, to the adsorption of the pesticide lindane (K_{ow} = 643 and K_{oc} = 1081 : Rao and Davidson, 1982) on a soil with an organic carbon content of 2% . It was assumed that adsorption of the pollutant could be described by linear isotherm at equilibrium ($N1 = 1$).

For any K_c value used in the simulations, Figure 1 presents the fraction of the bound pollutant, α_{hp}, obtained in solution at equilibrium for various dissolved humic substances concentrations, (H) (Eq. 3). If one assumes an average molecular weight and a fraction of organic carbon for humic substances of 2000 and 0.55, respectively, a DOC concentration of 1 mg/L would correspond to a $\approx 10^{-6}$ mol/L solution. Equation (3) was applied since the concentration of humic substances (10^{-5} mol/L) is much greater than the input pollutant concentration (10^{-7} mol/L): under such condition, the free humic substances concentration at equilibrium (H) approaches the total initial humic substances concentration.

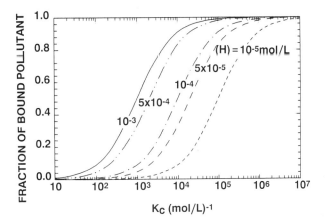

Figure 1.   Fraction of bound pollutant (α_{hp}) as a function of K_c.

A sensitivity analysis of the model was performed to determine the effect of variations in key physico-chemical parameters on calculated BTCs for hypothetical pollutants. Sets of simulations were carried out in order to study: 1) the effect of sorption kinetics (third site) in the three-site model; 2) the effect of pore-water velocity using the three-site model, and; 3) the effect of complexation using the first site model (instantaneous equilibrium model) and the two-site model (equilibrium-kinetic model). The effects of individual sorption or complexation parameters were examined by holding constant all other variables.

Effect of sorption kinetics and pore-water velocity in the three-site model

The three-site model (Sp1, Sp2 and Shp) simplifies to a typical two-site (Sp1 and Sp2) equilibrium-kinetic model (Selim et al., 1976; Rao et al., 1979) when no complexation of the pollutant occurs in solution. The influence of the second site on the BTC for a pollutant has been studied previously (Lafrance et al., 1987; 1989). In the present study, we considered the influence of the third site (Shp), with slow adsorption occurring on the second site. The effect of the magnitude of the k_3 and k_4 rate constants on the effluent curves is illustrated in Figures 2 and 3 for average pore-water velocities of 2 cm/h and 1 cm/h, respectively. The ratio $k_3/k_4 = 20$ was kept constant. The use of K = 0 (i.e., Sp1 sites unoccupied) allowed an evaluation of the significance of k_3 and k_4 without modification of the term R in Equation (6). The use of $K_c = 10^6$ implies that $\alpha_{hp} = 0.91$.

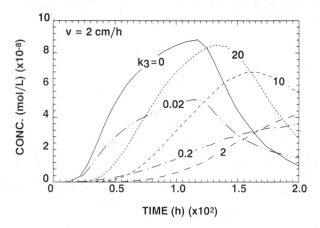

Figure 2. Effect of changes in the rate coefficients k_3 and k_4 in the three-site adsorption model. K = 0; $k_1 = 0.2$; $k_2 = 0.01$; $K_c = 10^6$; $k_3/k_4 = 20$; v = 2 cm/h.

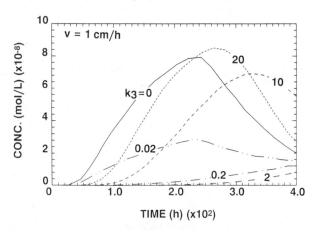

Figure 3. Effect of changes in the rate coefficients k_3 and k_4 in the three-site adsorption model. K = 0; $k_1 = 0.2$; $k_2 = 0.01$; $K_c = 10^6$; $k_3/k_4 = 20$; v = 1 cm/h.

For a pore-water velocity v = 2 cm/h, Figure 2 shows that an increase in the magnitude of both k_3 (from 0 to 0.2) and k_4 values causes a regular decrease of the breakthrough concentration result and a curve distribution moving to the right. Further increase in the sorption rate constants k_3 (from 2 to 20) and k_4 results in a decrease of the overall retardation process and in a shift to the left of the effluent distribution curves: the sorption process on the third site then approaches instantaneous equilibrium. The importance of the third site with respect to the two-site nonequilibrium adsorption model will be determined at one and the same time by the K_c value and by the relative importance of k_3 with respect to k_2. In fact, the adsorption of the complex on the third site diminishes the "carrier-effect" of K_c observed when using the two-site adsorption model.

When nonequilibrium sorption contributes to the retention process in soil, variations in the pore-water velocity may significantly affect the concentration distribution for a pollutant (Selim et al., 1976; Rao et al., 1979). Figure 3 presents the simulated BTCs obtained with the same variations in k_3 and k_4 rate constants, but with a pore-water reduced from 2 cm/h to 1 cm/h. In order to compare the resulting BTCs, the simulation time and the duration of injection have been chosen to keep the value of vt constant. Diminishing the pore-water velocity allows more time for the kinetically controlled sorption sites to interact, the result being a marked decrease of the breakthrough concentration for low values of the sorption rate constants ($k_3 < 2$). For higher values of the rate constants (e.g., $k_3 > 10$), the distribution curves approach the typical position of an instantaneous equilibrium and are not affected by the decrease of the pore-water velocity. The impact on BTCs of slow sorption kinetics for the free or the bound form of the pollutant will thus be dependent on the hydrodynamic conditions encountered in the subsurface.

Effect of complexation (K_c) in the two-site model (no adsorption of the complex)

The effect of variations in the K_c value was examined with respect to the instantaneous equilibrium adsorption model, i.e. first group Sp1 only (Figure 4), and to the equilibrium-kinetic adsorption model, i.e. first group Sp1 and second group Sp2 (Figure 5). With the equilibrium adsorption model (Figure 4), increasing the K_c value from 10^4 to 10^6 (i.e. decreasing α_p) markedly enhances the overall pollutant transport through a soil column. The fraction of the free form of the pollutant, α_p, being directly related to the retardation factor, a decrease in α_p causes a decrease in the residence time of the pollutant. A value of $K_c = 10^6$ implies nearly complete complexation of the pollutant in solution, which then adopts the behavior of a nonreactive solute. Similar variations in the K_c value in a two-site sorption model, i.e. no adsorption of the complex (Figure 5), result in both a decrease in the residence time and an increase in the maximum concentration of the effluent curves. This may be an important consideration when experimental BTCs obtained in the presence of dissolved non-adsorbable ligands are interpreted with nonequilibrium adsorption models.

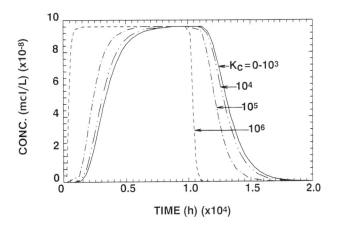

Figure 4. Effect of changes in the complexation constant (K_c) on the BTCs obtained with the instantaneous adsorption equilibrium model (first site). K = 20; $k_1 = 0$.

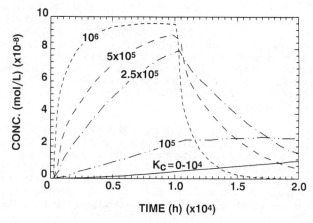

Figure 5. Effect of changes in the complexation constant (K_c) on the BTCs obtained with the two-site equilibrium-kinetic adsorption model. $K = 20$; $k_1 = 0.2$; $k_2 = 0.01$.

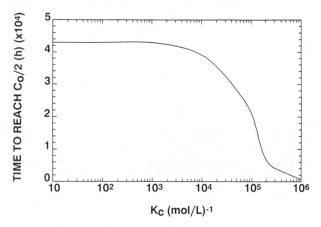

Figure 6. Time to reach $C_0/2$ from the BTCs obtained with the two-site model (Fig. 5) as a function of K_c.

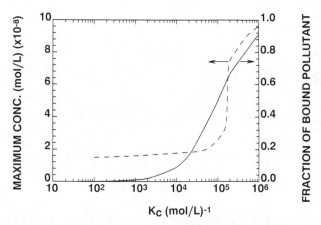

Figure 7. Effect of changes in K_c on the maximum concentration of the BTCs obtained with the two-site adsorption model (Fig. 5). The fraction of bound pollutant (α_{hp}) is also presented.

As shown in Figure 6, variations in K_c from 10^4 to 10^6 greatly affect the time required to reach 0.5 times the pollutant input concentration ($C_0/2$) as indicated by the BTCs obtained from the two-site sorption model (Figure 5). Figure 6 shows results obtained from the BTCs when injection of the pollutant was maintained throughout the simulation in order to reach the value $C_0/2$. On the other hand, Figure 7 shows that the variation in maximum breakthrough concentrations, as a function of K_c, is closely related to variations in the non-adsorbable bound fraction of the pollutant, α_{hp}. The "carrier-effect" of K_c could be reduced only by a higher adsorbability of the complex on the soil matrix.

Discussion on possible applications to chemical transport studies

In this study, all simulations were performed with values for physico-chemical parameters chosen to be representative of non-point source contamination of groundwater by HOC (e.g., pesticides in agricultural zones). Low HOC and DOC concentrations were assumed, such that the K_c value has to be greater than 10^5 to obtain more than 50% of the HOC solution concentration in bound form. Such values for K_c are known to apply for some pesticides (Carter and Suffet, 1983) or PAHs (McCarthy and Jimenez, 1985). According to McCarthy and Jimenez (1985), the values of DOC/water partition coefficients or K_p for some PAHs (K_p from 10^3 to 10^6) were closely related and of the same order of magnitude as the K_{ow} values for the compounds. The value of K_c (mol/L)$^{-1}$, as used in the model, could be obtained from K_p ([mol/g DOC]/[mol/g water]) by the relation: $K_c = K_p \times 10^{-3} \times f_{oc} \times$ molecular weight of DOM; where f_{oc} is the fraction of organic carbon in the DOM. From such a relation, the K_c value is expected to be nearly identical to that of K_p, if an average molecular weight of 2000 for the DOM and a fraction of organic carbon of 0.50 are assumed. With reference to the relationship between K_{ow} and K_p, one may expect that the transport of HOC with K_{ow} values greater than 10^5 will be significantly affected by DOM concentrations encountered in soil-water systems, provided the HOC-DOM complex is poorly retained on the soil matrix.

With respect to our knowledge concerning the extent of binding between HOC and DOM, it is obvious that such interactions must be considered in any conceptual model designed to simulate the transport of HOC towards the aquifer. However, many unknown natural conditions or phenomena still limit the application of such deterministic models. In particular, the variability of the concentration and reactivity of DOM may impose large variations in the extent of binding of HOC (Carter and Suffet, 1983; Gauthier et al., 1987). Despite the current lack of data representative of field situations, the present model may serve as a reference basis in describing HOC transport in soil column studies involving natural DOM.

CONCLUSION

To predict the extent to which binding of hydrophobic compounds by natural dissolved organic matter could affect their transport in the subsurface, a three-site complexation-adsorption model was developed. From an analysis of the sensitivity of the proposed model to variations in selected sorption and complexation parameters, the following conclusions are drawn:

1- If one assumes slow kinetics for the sorption on soil of the bound form of a contaminant, the relative impact of adsorption rate contant on the mobility of the contaminant will be determined at one and the same time by the extent of binding in solution and by the pore-water velocity in the subsurface. Retention of the complex on a non-competitive sorbing site would attenuate the possible enhancement of contaminant transport attributable to complexation. However, the pore-water velocity would have to be sufficiently low to allowed the kinetic site to interact.

2- The formation of non-adsorbable complexes greatly affects the mobility of a contaminant in soil. In a two-site equilibrium-kinetic model for the sorption of the free form of the contaminant, complexation enhances the breakthrough of the contaminant in a soil column (effect on the first site) and also increases the maximum concentration of the resulting effluent curves (effect on the two sites). The maximum concentration is closely related to the bound fraction of the contaminant, this fraction being a function of both the complexation constant, K_c, and the dissolved organic matter concentration.

3- For natural dissolved organic matter concentrations encountered in soil-water systems (around 10 mg/L DOC as humic substances), the suggested model predicts a significant effect of complexation on the mobility of contaminants in the micromolar concentration range that present K_c values greater than 10^5, provided that the resultant complexes are poorly retained on the soil matrix. Such values for K_c are expected to apply to various potential contaminants.

ACKNOWLEDGMENTS

The authors express their appreciation to the Donner Canadian Foundation and to the Natural Sciences and Engineering Research Council of Canada for financial support.

REFERENCES

Bengtsson, G., Enfield, C.G. and Lindqvist, R. (1987). Macromolecules facilitate the transport of trace organics. *Sci. Total Environ.*, 67, 159-164.

Caron, G. and Suffet, I.H. (1987). The binding of tetrachlorobiphenyl to dissolved organic carbon in sediment interstitial waters. Paper presented before the 193rd National Meeting, Am. Chem. Soc., Div. Environ. Chem., Denver, CO, April 5-10, 1987, Preprints Extended Abstracts, 27 (1), 193-194.

Carter, C.W. and Suffet, I.H. (1983). Interactions between dissolved humic and fulvic acids and pollutants in aquatic environments. In: *Fate of Chemicals in the Environment. Compartmental and Multimedia Models for Predictions*, R.L. Swann and A. Eschenroeder (Eds.), Am. Chem. Soc. Symposium Series, No. 225, Washington, DC, pp. 215-229.

Enfield, C.G. and Bengtsson, G. (1988). Macromolecular transport of hydrophobic contaminants in aqueous environments. *Ground Water*, 26, 64-70.

Gauthier, T.D., Shane, E.C., Guerin, W.F., Seitz, W.R. and Grant, C.L. (1986). Fluorescence quenching method for determining equilibrium constants for polycyclic aromatic hydrocarbons binding to dissolved humic materials. *Environ. Sci. Technol.*, 20, 1162-1166.

Gauthier, T.D., Booth, K.A., Grant, C.L. and Seitz, W.R. (1987). Fluorescence quenching studies of interactions between polynuclear aromatic hydrocarbons and humic materials. Paper presented before the 193rd National Meeting, Am. Chem. Soc., Div. Environ. Chem., Denver, CO, April 5-10, 1987, Preprints Extended Abstracts, 27 (1), 243-245.

Hassett, J.P. and Milicic, E. (1985). Determination of equilibrium and rate constants for binding of a polychlorinated biphenyl congener by dissolved humic substances. *Environ. Sci. Technol.*, 19, 638-643.

Keoleian, G.A. and Curl, R.L. (1987). Effects of humic acid on the adsorption of tetrachlorobiphenyl by kaolinite. Paper presented before the 193rd National Meeting, Am. Chem. Soc., Div. Environ. Chem., Denver, CO, April 5-10, 1987, Preprints Extended Abstracts, 27 (1), 183-185.

Lafrance, P., Banton, O., Villeneuve, J.P. and Campbell, P.G.C. (1987). Simulation with a complexation-adsorption model of possible effects of dissolved humic substances on solute transport in soils. Paper presented before the National Water Well Association Conference: "Solving Groundwater Problems with Models", February 10-12, 1987, Denver, CO. Proceedings: Vol. 1, 69-87.

Lafrance, P., Banton, O., Campbell, P.G.C. and Villeneuve, J.P. (1989). Modeling solute transport in soils in the presence of dissolved humic substances. *Sci. Total Environ.*, in press.

McCarthy, J.F. and Jimenez, B.D. (1985). Interactions between polycyclic aromatic hydrocarbons and dissolved humic material: binding and dissociation. *Environ. Sci. Technol.*, 19, 1072-1076.

Ogata, A. and Banks, R.B. (1961). A solution of the differential equation for longitudinal dispersion in porous media. Prof. Pap. 411-A, U.S. Geol. Survey, 7 p.

Paxéus, N., Allard, B., Olofsson, U. and Bengtsson, M. (1985). Humic substances in ground waters. In: *Scientific Basis for Nuclear Waste Management IX*, L.O. Werme (Ed.), Materials Research Society Symposia Proceedings, Stockholm, Sweden, September 9-11, 1985, Vol. 50, 525-532.

Rao, P.S.C., Davidson, J.M., Jessup, R.E. and Selim, H.M. (1979). Evaluation of conceptual models for describing nonequilibrium adsorption-desorption of pesticides during steady-flow in soils. *Soil Sci. Soc. Am. J.*, 43, 307-311.

Rao, P.S.C. and Davidson, J.M. (1982). *Retention and transformation of selected pesticides and phosphorus in soil-water systems: a critical review*. Report EPA 600/3-82-060, May 1982, 321 p.

Selim, H.M., Davidson, J.M. and Mansell, R.S. (1976). Evaluation of a two-site adsorption-desorption model for describing solute transport in soils. In: Proceedings, Summer Computer Simulation Conference, July 12-14, 1976, Nat. Sci. Found., Washington, DC, pp. 444-448.

Thurman, E.M. (1985). *Organic Geochemistry of Natural Waters*, M. Nijhoff and W. Junk (Pub.), Dordrecht, The Netherlands, 497 p.

Thurman, E.M. and Malcolm, R.L. (1983). Humic substances in groundwater. Paper presented before the 186th National Meeting, Am. Chem. Soc., Div. Environ. Chem., Washington, DC, September, 1983, Preprints Extended Abstracts, 23 (2), 242-244.

Thurman, E.M., Wershaw, R.L., Malcolm, R.L. and Pinckney, D.J. (1982). Molecular size of aquatic humic substances. *Org. Geochem.*, 4, 27-35.

THE EFFECT OF EXCHANGEABLE CATIONS ON THE PERMEABILITY OF A BENTONITE TO BE USED IN A STABILIZATION POND LINER

P. M. Büchler

Department of Chemical Engineering, Polytechnic School, São Paulo University, P.O. Box 8.174, São Paulo, SP, 01000 Brazil

ABSTRACT

The organophilic nature of bentonites exchanged with quaternary ammonium cations is used in sanitary engineering for the adsorption of organic pollutants. This paper deals with five different quaternary ammonium cations: tetramethylammonium, trimethylstearylammonium (C_{18}), dimethylbenzyllaurylammonium (C_{12}), trimethylpalmitylammonium (C_{16}) and dimethyldistearylammonium. A Brazilian bentonite was treated with the above cations and the adsorption of vinasse organics was measured through the total organic carbon present in solution. The results show that tetramethylammonium cation is the most effective of those tested to make sodium bentonite more organophilic and the behaviour follows a Freundlich isotherm. If the isotherms are plotted in milliequivalents of the cation over the weight of the sodium bentonite the present experiments did not show an appreciable difference in the quantity adsorbed. Therefore, if cost is a determining factor, low molecular weight cations should be chosen. The modified bentonites were characterized by the X-ray diffraction patterns. For high molecular weight cations the interlamelar spacing is close to 18 Å but for tetramethylammonium it is 13.5 Å. In any case the replacement of sodium by a quaternary ammonium cation increases the capacity of the clay to adsorb organic molecules.

KEYWORDS

Organophilic bentonites; organics adsorption; vinasse treatment; fatty acids quaternary ammonium salts; subsurface contaminants attenuation.

INTRODUCTION

Previous studies had shown the effectiveness of organophilic clays in the adsorption of organic components of vinasse (Büchler, 1987). Vinasse is the residue from distillation of fermentation ethyl alcohol. Since it is stored in ponds prior to use as fertilizer, cattle feed supplement or anaerobic treatment it may have a dangerous effect in the subsurface environment. Sodium bentonites are used as sealing agents for water ponds but because of calcium and magnesium cations present in vinasse the permeability of sodium bentonite liners tends to decrease with time (Grim, 1968). When the sodium cation is replaced by a quaternary ammonium cation the bentonite tends to deflocculate and it will be no longer impermeable. The association of sodium bentonite and an organophilic bentonite can be used in storage ponds to attenuate the infiltration of organic pollutants into the underground water. Some researchers (McBride et al.,1985) had shown that low molecular weight quaternary ammonium cations are more effective to increase the adsorption capacity of sodium bentonites than high molecular weight cations. Quaternary ammonium salts made of fatty acids can easily be purchased because they are used as fabric softeners. Early experiments dealing with adsorption of organics on organophilic clays were performed with low molecular weight quaternary ammonium cations (Barrer, 1955) but more recently larger molecules are being tested (Boyd et al.,1988). Working with aqueous solutions of phenol McBride et al (1985) found 27% removal for tetramethylammonium substitution and 8% for hexadecyltrimethylammonium. The mechanism explained for adsorption of bentonites exchanged with quaternary ammonium cations by Büchler and Perry (1986) is: the sodium cation present in most bentonites is easily hydrated by six molecules of water and that is the main reason to explain the intense swelling of sodium bentonites. This means that sodium bentonites are extremely hydrophilic. But when sodium is replaced by a quaternary ammonium cation the hydrophobic nature of the new cation makes the clay surface become hydrophobic and therefore organophilic.

EXPERIMENTAL PROCEDURE

The sodium bentonite used in the following experiments came from a Brazilian mine located in the state of Paraiba. Since it is a calcium bentonite it was previously treated with an aqueous solution of sodium carbonate to become a sodium bentonite similar to the Wyoming bentonite.The sodium cation is exchanged with the quaternary ammonium cation by making a suspension in the proportion of 3 grams of the clay and 100 ml of a 1 molar solution of the salt. The exchanged clays are characterized by the X-ray diffraction method. The equipment supplied by Phillips Company from Holland works with the k-alpha radiation of copper.The suspension is mechanically stirred for 24 hours. In order to assure a substitution reaction as close as possible to completion the the particles of sodium bentonite used are smaller than 2 μm. This is achieved using only the bentonite which remains in suspension after 24 hours that it is mixed with water. If this procedure is not carefully followed the final results may vary appreciably. The sodium bentonite which in pure water swells up to 15 times its original volume will settle as a consequence of the reaction with the quaternary ammonium salt. After the reaction is completed the suspension is centrifuged for 15 minutes at 1,000 rpm and the supernatant is discarded. The exchanged clay is then washed with water until a drop of a 1 molar solution of silver nitrate does not precipitate any silver chloride. The clay is dried at $80°C$ after centrifugation until constant weight. The adsorption experiments with the modified bentonites are performed with a mixture of 500 mg of the clay and 100 ml of diluted vinasse solution. Prior to dilution the vinasse is filtered in a millipore filter to remove the suspended matter. The suspension is stirred in a constant temperature bath for 24 hours at $30°C$ and then centrifuged for 15 minutes at 1,000 rpm. The amount of organics adsorbed at the vinasse solution is measured with the assistance of a total organic carbon analyser. A sample of 40 μl is injected in the apparatus. The organic carbon in solution reacts with a sodium persulphate solution and catalytic ultra violet light. The carbonic gas thus formed is measured by an infrared analyser. The result is displayed in a screen and printed.

RESULTS AND DISCUSSION

The X-ray diffractograms of the modified clays are shown in Figure 1. The basal spacing, i.e., the space between the layers of silica and alumina in the crystalline structure of the clay is 13.5 Å when sodium is replaced by the low molecular weight quaternary ammonium cation (tetramethylammonium). For the higher molecular weight cations the distances are between 18 and 19 Å. Figure 2 shows the adsorption isotherms of vinasse solutions on the surface of modified clays at 30°C. All lines follow a Freundlich-type isotherm. The linear regression for the curve fitting indicates correlation coefficients above 90% in all cases. The isotherms show a clear influence of the quaternary ammonium salt molecular weight. The higher adsorption occurs for C_4 ; the lower for C_{38} and the intermediate values were approximately the same for number of carbon atoms close to C_{20}.

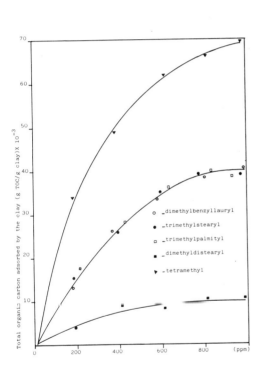

Fig. 1 Adsorption of vinasse on modified bentonites at 30°C

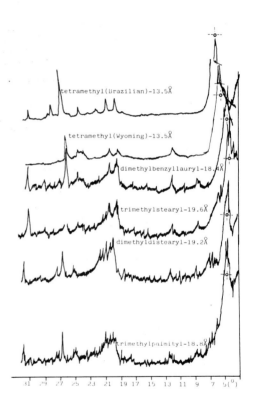

Fig. 2 X-ray diffractograms of the modified bentonites

CONCLUSION

If cost is competitive low molecular weight quaternary ammonium cations should be chosen. In the present experiments only tetramethylammonium cation was used. McBride et al. (1985) shows that for tetraethylammonium the adsorption of phenol from aqueous solutions drops from 27% for Wyoming sodium bentonite exchanged with tetramethylammonium cation to 5%.

ACKNOWLEDGEMENT

This research was made possible thanks to grants from: The Fulbright Commission, The British Council, The Brazilian Ministry of Education (Capes), The Brazilian Research Coucil (CNPq), The São Paulo State Research Agency (Fapesp) and The São Paulo University Commission for International Cooperation (CCInt). The author wants also to thank Professor Roger Perry from Imperial College in London and Dr. Walter J. Weber, Jr. from The Michigan University for permission to use both universities facilities.

REFERENCES

Barrer, R.M. and MacLeod, D.M. (1955). Activation of montmorillonite by ion exchange and sorption complexes of tetraalkylammonium montmorillonites. Trans. Faraday Soc., 51, 1290-1300.

Boyd, S.A., Mortland, M.M., Pinnavaia, T.J. and Weber, Jr., W.J. (1988). Use of modified clays for the sorption and catalytic degradation of VOC's. Research Proposal, The Michigan University and The Michigan State University

Büchler, P.M. (1987). The attenuation of underground water contamination in a vinasse clay-lined pond. Wat. Sci. Tech., 19 (12), 183-185.

Büchler, P.M., Warren, D., Clark, A.I. and Perry, R. (1987). The use of clay liners in the attenuation of the organic load of vinasse in developing countries. Int. Conf. Chem. in the Environ., 715-724, Lisbon (1986).

Grim, R.E. (1968). Clay Mineralogy, 2nd ed., McGraw-Hill, New York.

McBride, M.B., Pinnavaia, T.J. and Mortland M.M. (1985). Adsorption of aromatic molecules by clays in aqueous solutions. Advan. Environ. Sci. Technol., 8 (1), 145-154.

MOBILITY OF SOLUBLE AND NON-SOLUBLE HYDROCARBONS IN CONTAMINATED AQUIFER

J. Ducreux*, C. Bocard*, P. Muntzer**, O. Razakarisoa** and L. Zilliox**

*Institut Français du Pétrole, B.P. 311, 92506 Rueil-Malmaison cédex, France
**Institut de Mécanique des Fluides, Université Louis Pasteur, URA CNRS 854, 2, rue Boussingault, 67083 Strasbourg cédex, France

ABSTRACT

After the contamination of an aquifer by petroleum products, the residual oil trapped is a constant source of pollution by the entrainment of the most soluble hydrocarbons. By studying the exchanges of residual hydrocarbons between oil-water-air and soil, we pointed out that the liquid/gas exchange is the major factor of retention of soluble alkanes masking the adsorbing materials effects. For the soluble aromatic hydrocarbons, the main phenomenon observed is the liquid/solid exchange. The role of residual air is no more preponderant.

The residual contamination of the vadose zone thus plays a preponderant role in the long-term pollution of a groundwater table. It is thus imperative to implement methods to prevent such harmful effects. The use of surfactants, by lowering the oil/water interfacial tension seems to be a new and effective method. Their adsorption into a natural matrix was studied with different porous substrates (sand, sand/silt). Their retention on sand is poor, but it increases with silt content. This is mainly due to a cationic exchange (Ca^{2+}/Na^{+}). In order to avoid this phenomenon a salt preflush by a 10 g/l Na Cl solution is effective. That allows a gas-oil recovery enhancement by reducing loss of surfactant in soil. Moreover, a surfactant partition between oil and water is underscored. A better understanding of these parameters would lead to the optimizing of the enhanced drainage technique for recovering residual oil trapped in an aquifer.

KEYWORDS

Aquifer, hydrocarbons, exchange processes, anionic surfactants, adsorption, oil recovery techniques.

INTRODUCTION

Contamination of underground by oil can result from accidental spills or leaking underground storages. Even when mobile oil, which has reached the water table, has been removed by pumping, residual trapped oil can be a long-term source of water contamination by soluble hydrocarbons which should be neutralized (Béraud et al., 1989). Biodegradation and drainage enhanced by surfactants have been considered, but other techniques such as venting can be implemented in the case of light oils, especially gasoline. Hence, it was necessary to get a better understanding of the behaviour of light hydrocarbons, especially with regard to the exchange processes between solid, liquid and gas phases.

The mobility of soluble alkanes and aromatic hydrocarbons has been conducted at the laboratory with experimental dual-column model to underscore the exchange mechanisms between liquid/solid and liquid/gas phases in vadose and saturated zones.
The experimental approach of these physico-chemical exchanges can give answers to questions such as :

- What is the role of residual air compared with that of solid material, on the exchange possibility with the water charged with dissolved hydrocarbons ?
- Is the observed exchange phenomenon linked to the nature and the chemical properties of soluble hydrocarbons ?

In order to fight against the immobilized residual oil contamination an enhanced drainage technique by using anionic surfactant was proposed. To improve the effectiveness of these products when they are injected in adsorbing porous media, the adsorption parameter was examined with experimental model of the column-type, and a salt preflush technique was developped to decrease the loss on such matrixes.

EXPERIMENTAL

Mobility of soluble hydrocarbons

The main mechanism studied was the influence of the porous medium air content. The variation of the presence of residual air was obtained according to the filling of the column : "dry filling", favours the presence of residual air, "filling by sedimentation" can reduce or partially avoid it.

The porous medium is constituted of natural quartz sand or a mixture of soil and quartz sand. The soil contained an average of 3 % of organic materials, 7 % of clay minerals and 26 % of limestone.

Continuous injections of hydrocarbon solutions made of a binary mixture of alkanes (cyclohexane + 2,3-dimethylbutane), or a mixture of four aromatics (toluene + p-xylene + 1,2,3-trimethylbenzene + 1-methylnaphthalene),were carried out through the column.

The study of exchange mechanism between liquid/solid and liquid/gas phases has been conducted at the laboratory with experimental models of the dual column type. The experiment consisted in getting water charged with dissolved hydrocarbons (source : first column, concentration C_o), flowed through a second column which was not contaminated at the beginning. This model was described elsewhere (Razakarisoa et al., 1989).

The analysis and the evolution of the contents of soluble hydrocarbons in water coming out from the second column (concentration C) allowed to apprehend the observed exchange phenomenon.

Mobility of non-soluble hydrocarbons by using surfactant application

A tracer (potassium iodide) was used to determine the accessible pore volume (Vp). Its concentration was measured by UV detector.

Materials. The materials used in the experiments were selected to simulate those encountered in a water-table aquifer : sand and sand/silt mixtures.
The grain-size range of sand was relatively narrow, thus 79 % got a range between 297 and 420 μm (0.0116 to 0.0165 inch).

The silt "MOTELLE" was provided by INRA (Institut National de Recherche Agronomique - France). Its granulometric and chemical characteristics are shown in Table 1.

TABLE 1   Granulometric and chemical characteristics of silt "MOTELLE"

Granulometric Analysis		Chemical Analysis	
Clay	24.2	Total limestone	8.1
Fine silt	26.4	Exchangeable calcium	15.7
Coarse silt	31.2	Exchangeable magnesium	29.1
Fine sand	15.8	Exchangeable potassium	1.7
Coarse sand	2.4	Exchangeable sodium	0.5
		Exchangeable aluminium	0.5

An organic fluid (gas-oil) that is lighter than water and also immiscible was of interest because it is transported and stored in large quantities and can lead to contamination of the subsurface.

The surfactant used in the experiments was a proprietary product. It was a sodic anionic compound.

Columns. A 2.5 cm (0.995 inch) diameter glass column (Pharmacia) was fitted with nylon screens, and was connected to a syringe pump (P500- Pharmacia). The filling method was "by sedimentation" ; the weight of material was 62.5 g. The packed sand column had the following physical properties : water-saturated hydraulic conductivity $(Ks) \simeq 6 \times 10^{-5}$ m/s (1.9 x 10^{-4} ft/s) ; porosity $(n) \simeq 0.38$ (38 %) ; bulk density $(\rho_b) \simeq 1640$ kg/m^3 (10^2.4 lb/ft^3). When we used the sand/silt mixtures (5 or 10 % silt), prior to the filling, a calcium chloride solution was added (0.5 %) and vacuum dried, in order to have an homogeneous packing. Therefore, the porous medium was washed by water ($\simeq 100$ (Vp) pore volume) prior to the test. The temperature was 20°C. The liquid flow rate was 21 ml/h. The outflow was connected to a fraction collector (FRAC 100 - Pharmacia). The volume of each fraction was 2 ml.

Adsorption of surfactants on sand/silt columns. The surfactant used was dissolved (Co) in salt water (NaCl = 2 g/l). The total injected volume was 150 ml. Quantification of surfactant in the outflow (C) was accomplished by using the Hyamine 1622 titration (NFT 73-258).

Enhanced gas-oil recovery tests (preflush techniques). In these experiments the oil was in residual saturation (Wo) prior to the surfactant solution application.

These columns are representative of the capillary fringe, where the residual oil remains entrapped after fluctuation of the water table.

The conditions of surfactant injection were the same as those used in adsorption tests on sand/silt columns.

The pre-flush step, using a sodium chloride solution (2 g/l or 10 g/l), was accomplished prior to oil injection. Its volume was about 41 pore volumes i.e. 500 ml. The goal of this treatment was to change the ionic nature of the silt surface by a cationic exchange (Ca^{2+} - silt/Na^+ - salt solution).

RESULTS AND DISCUSSION

Mobility of soluble hydrocarbons

The different elution graphs in response to the continuous injection of alkanes show a higher retention and a more important delay when the presence of residual air is favoured in the second column whatever the porous support which is used.

When the presence of residual air is minimized, fixation of hydrocarbons by the solid material is observed and is enhanced by the proportion of soil (Fig. 1).

When the dry filling of the second column is adopted, no significant differences appear between the elution graphs corresponding to the two tests using the matrix with 10 % or 20 % of soil. So, for the soluble alkanes, when the content of residual air confined in the porous medium becomes great, the role of the gaseous phase, relative to the retention of soluble hydrocarbons, masks those of solid material.

These results reveal the preponderant role played by residual air in the case of alkanes ; the part of the solid material is negligible for a natural quartz sand ; the liquid/gas exchange is the major factor of retention (Rasolofoniaina et al., 1988).

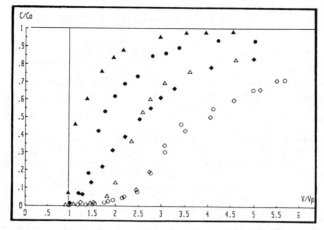

Fig. 1. Elution graphs in response to the continuous injection of soluble hydrocarbons made of <u>binary mixture of alkanes</u> (cyclohexane + 2,3-dimethylbutane).

For the soluble aromatic hydrocarbons, the evolution of the concentrations shows identical patterns of the pollutant whatever the air content of the matrixes is (Razakarisoa et al., 1989). This phenomenon is observed for each of the experiments using the three types of porous support (Fig. 2) :
- For a natural quartz sand, they practically behave as a tracer
- For a porous support with 10 % or 20 % of soil, the elution curves show a fixation and a delay due to the retention capacity of the matrix.

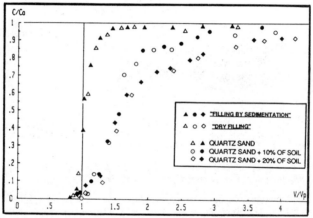

Fig. 2. Elution graphs in response to the continuous injection of soluble hydrocarbons made of <u>aromatic mixture</u> (Toluene + p-xylene + 1,2,3-trimethylbenzene + 1-methylnaphthalene)

Then, the role played by the residual air is no more preponderant in the case of the aromatics. The main phenomenon observed is the liquid/solid exchange.

The transfer of soluble hydrocarbons in the gaseous phase varies according to the type and nature of hydrocarbons. If the exchange between the water charged with dissolved hydrocarbons and the residual air is instantaneous (Johnson et al., 1987), equilibrium concentrations can be calculated by Henry's law : $C_a = H^* C_w$ where C_a and C_w are respectively the pollutant concentration in the air and in the water, $H^* = H/RT^a$ is computed according to Schwarzenbach's method (1981) ($H = P_v/S$), where P_v is the vapour pressure (Weast, 1978), S the solubility, R the constant of perfect gas and T the temperature. Values of the Henry's constant for the different hydrocarbons used are presented (Table 2). The value assigned to H^* allows to determine if the hydrocarbon is more or less retained by the gaseous phase (in relative value).

TABLE 2

Product	Vapour pressure at 20°C (mm Hg)	Solubility (mg/l)	H*
Hexane	126.6	12.3	0.429
2,3-dimethylbutane	193.0	20.2	0.399
Cyclohexane	79.6	56.4	0.058
Toluene	22.4	520	0.002
p-xylene	5.9	156	0.002
1,2,3-trimethylbenzene	2.8	75	0.002

Hydrocarbon solubility is not a sufficient criterion to explain the retention of soluble hydrocarbons. The more H* is small, the more the transfer of the hydrocarbon in the air is difficult. The tendency is confirmed by our experiments.

Mobility of non-soluble hydrocarbons by using surfactant application

The in situ treatment of hydrophobic organics (gas-oil, gasoline ...) can sometimes be accomplished by water injection/recovery systems using surfactants, followed by final treatment above ground. This approach is comparable to that used in enhanced oil recovery techniques.

The enhanced oil recovery (EOR) techniques displace residual oil by reducing the interfacial energy between the phases (oil, water) and minimize the viscosity of oil-water emulsions. Both of these characteristics are needed to promote flow of residual oil in the presence of water (saturated zone) and air (vadose zone) in a porous medium.

The effects of residual oil on capillary pressure (ΔP) can be explained qualitatively by a highly simplified model (Fig. 3).

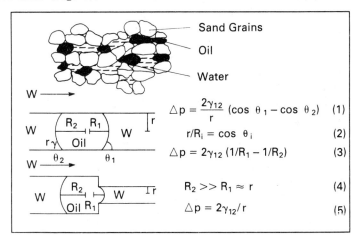

Δp = Capillary pressure (bars/meter) γ_{12} = interfacial tension (mN/m)

Fig. 3. Capillary pressure in a pore and in a pore constriction (from : Balzer D. 1988)

Inferring from equation (5), there are two possibilities to decrease the capillary pressure, either by increasing r, which is difficult in an aquifer, or by decreasing γ_{12} (interfacial tension). This one can be reduced by using surfactant solutions. Very low interfacial tensions are often observed after adding sodium chloride to the surfactants. With the surfactant used in our experiments, the interfacial tension between gas-oil and water decreased from 0.202 mN/m (active surfactant concentration = 0.5 %) to 0.072 mN/m, when 2 g/l salt solution was used (spinning drop technique measurement).

Texas Research Institute conducted studies on surfactant-enhanced gasoline recovery in sand column (1979), and in a large-scale model sand aquifer (1985) and showed a certain effectiveness of these additives, mainly with nonionic/anionic surfactants mixtures.

As a general rule, the effectiveness of this treatment on site appeared limited by soil adsorption of surfactants due to the presence of clay, silt ...

Study of the adsorption or retention of anionic surfactants during their migration in different porous media. Minerals which are likely encountered in a soil can react with the anionic surfactants and generate adsorption, precipitation and chimisorption processes. Clays are mainly involved in these phenomena.

At first, the elution of a sodic anionic surfactant was studied by using several packed columns (sand and sand/silt mixtures), then several ways were conducted in order to decrease the retardation of surfactant elution in silty porous media.

No oil calcic columns tests. Figure 4 displays the elution curves of sodic anionic surfactant through sand/silt columns and control test column (sand). It appears that the surfactant migration is linked to silt content. The surfactant retention in relative pore volume units in packed columns was respectively 1.2 and 1.6 for 5 % silt and 10 % silt. That demonstrates the retardation effect of clay in silt for such ionic surfactants. According to Shweich (1984) and Zundel and Siffert (1984) this retention is due to the calcium-sodium exchange process. Ion exchange takes place with clays.

The "motelle" silt used in the experiments contained an average of 23 % clay materials and 15.7 % exchangeable calcium ions (Table 1). When a sodium salt, as a sodic surfactant was used, an ion exchange was developed. In that case, a classical cationic exchange between the ion-counters Na^+ from surfactant and the exchangeable cations from clay (K^+, Ca^{2+}, Mg^{2+}) took place. In that step the retention is weak. It would correspond to either a neutralisation of clay positive sites or a chimisorption according to the following scheme :

$$\left[xNa^+, yCa^{2+} \right]_{clay} + 2\epsilon Na^+ \rightleftharpoons \left[(x+2\epsilon)Na^+, (y-\epsilon)Ca^{2+} \right]_{clay} + \epsilon Ca^{2+}$$

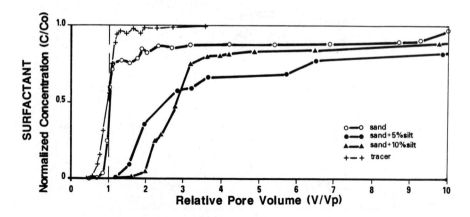

Fig. 4. Influence of the silt content on the surfactant retention

Therefore the calcium is virtually extracted from the entire mineral. The precipitation of the anionic surfactant as a calcic salt happens, mainly, just below its Critical Micellar Concentration (CMC) :

$$2 T^- + \epsilon Ca^{2+} \longrightarrow \epsilon Ca\, T_2 \text{ (solid)}$$

(T^- corresponding to anionic group from surfactant).

The CMC of surfactant used in the experiments was about 700 mg/l (1.58 meq/l) at 25°C. A retardation of surfactant elution was observed with this range of concentrations and corresponding to the relative pore volumes (V/VP) : 1.5 and 2.0. In these cases, the surfactant retention on calcic clay was strong. Above the CMC, a regular decrease of calcic surfactant was observed up to normalized concentration levels (C/Co = 0.8) lower than that of control test. This result can be explained, as suggested by Zundel et al. (1984), by a transfer of calcium in the micelles, according to the scheme :

$$T^-, nNa^+ + \epsilon Ca^{2+} \rightleftharpoons T^- \left[(n-2\epsilon) Na^+, \epsilon Ca^{2+}\right] + 2 \epsilon Na^+$$
(sodic micelles) (mixt micelles)

An equilibrium exists between the precipitation of calcium salt (CaT_2) and its dissolution. Therefore, the slight loss of surfactant observed is due to an incomplete dissolution of precipitated surfactant. Bazin and Defives (1984) attributed the retention of surfactant to its precipitation because of aluminium ions coming from the dissolution of kaolinite. Early investigators (Hill and Lake, 1977 ; Glover et al., 1978) thought that retention could be due to formation of surface complexes with Ca^{2+} and Mg^{2+} brought by the clay.

In general, only the exchangeable cations seem to be responsible of the retention of anionic surfactant on the clay or silt, with a possible chimisorption.

No oil sodic column tests : effect of the preflush treatment by sodium chloride solution on the retention of sodic anionic surfactant. In clean up process using surfactants, it is imperative to get an adsorption rate as low as possible, to improve its effectiveness and decrease the cost of treatment.

Considering the cation exchange between the injected anionic surfactant solution and silt, the effect of a preflush by a sodium chloride solution was studied to enhance a best surfactant recovery and decrease its retention.

Two concentrations of salt were used : 2 and 10 g/l (respectively 0.034 and 0.17 mole/l). Figure 5 displays the restitution of the sodic anionic surfactant during its elution through 5 % and 10 % silt packed columns, with and no preflush.

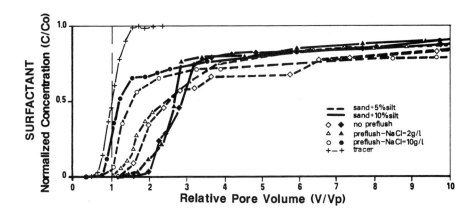

Fig. 5. Influence of salt preflush on the surfactant retention

In no preflush tests, the retention of surfactant was observed. A 2 g/l preflush salt solution had only a minimal effect on retention of surfactant, whereas a 10 g/l salt solution strongly decreased that retention. With a 10 % silt column, the elution of surfactant was even comparable with sand column test (Figure 4) considered with no adsorption or precipitation phenomena. It is probably due to the modification of cationic aspect of silt. When no preflush is applied the silt is in calcic form, with a 10 g/l sodium chloride solution, as it has been explained previously, the clay becomes sodic by cationic exchange (Ca^{2+}/Na^+) : the sodium gathers to the clay, releasing calcium ions. Following an efficient preflush (10 g/l) there is no possibility of cationic exchange between silt and surfactant. Therefore, the precipitation and associated retention of surfactant is avoided. As with no oil calcic column tests and for V/Vp value over three, the release of surfactant is satisfactory and its loss is depleted.

Enhanced gas-oil recovery by using surfactant application and preflush technique : oil mobilizing

Sand column, control test. Figure 6 displays surfactant release and gas-oil recovery in two tests : no oil control and oil test. In no oil control test, the surfactant behaved as a tracer except the values on step. There was no retention and its release was correct.

In the oil test, the oil was trapped in residual saturation. Following surfactant washdown there was an immediate oil mobilizing, and effective gas-oil removal. After only two Vp surfactant solution injections, more than 60 % of the volume spilled was removed as raw oil. 4.5 Vp were sufficient to remove 83 % of the residual oil. The final yield was 86.7 %. The amounts of dissolved or emulsified oil in effluent are much lower than that raw oil mobilized (yield = 4.6 %). So, the oil mobilized by using such a clean-up process, will be easily recoverable. This figure illustrates the effectiveness of surfactant washdown on mobilizing gas-oil entrapped in sandy aquifer. The plot illustrating the surfactant relative concentrations, in effluent, shows a noticeable difference compared to no oil control test curve, mainly at the beginning of surfactant washdown. The adsorption of surfactant on sand being weak, that difference is probably due to a reversible partition of anionic surfactant between two phases (gas-oil and water). That effect corresponds to the raw oil mobilizing and is in part due to the lipophilic character of the surfactant. When amounts of residual oil in porous medium are poor, its hydrophilic character becomes preponderant. Therefore, the surfactant is removed with the aqueous effluent.

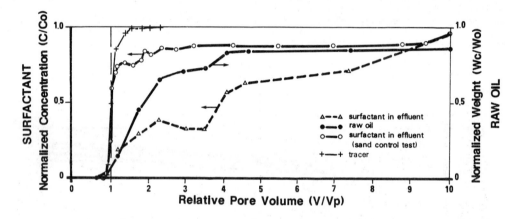

Fig. 6. Sand column, control test - Surfactant effectiveness on the gas-oil recovery. Effect of the surfactant partition between water and oil

Sand/silt column tests. The experiments illustrate the effectiveness of surfactant washdowns on enhanced gas-oil recovery according to the modification of the silt mixed with sand, by using salt preflush techniques. The silt content in sand/silt columns was 5 % (w/w).

Figure 7 displays cumulative curves of total raw gas-oil removed (Wc/Wo), and surfactant normalized concentrations (C/Co) in the effluents during each test : no preflush, salt preflush (2 and 10 g/l) control test on sand column. There are no noticeable differences between no preflush and 2 g/l (NaCl) preflush tests. These surfactant washdowns had a slow effect in mobilizing gas-oil. However, after 10 Vp surfactant solution injected the yields were respectively 92 and 82 %. The ineffectiveness of the 2 g/l (NaCl) preflush is probably due to too low sodium cations concentration to allow an efficient treatment by cations exchange. On the other hand, the 10 g/l salt solution preflush provided a best effectiveness in gas-oil removal. In that test, less surfactant was used to get the same yields.

Table 3 summarizes the results of effectiveness for each test. It illustrates the improvement brought by concentrated salt preflush (NaCl = 10 g/l) in mobilizing gas-oil entrapped in a silty soil. On site, that will allow to apply surfactant flooding on a shorter period of time, and decrease the surfactant treatment cost.

TABLE 3 Preflush effectiveness on raw-oil recovery

Relative pore volume	Effectiveness % of the oil entrapped			
V/Vp	no preflush	preflush (NaCl)		Sand control test
		2 g/l	10 g/l	
1	0.1	0	6	16
2	6	2	30	62
4	31	20	54	83
6	47	44	65	85
10	92	82	77	86.7

Noticeable differences appear in the release of surfactant in effluent between the tests established with or no oil conditions (Figure 7 and 5). The data curves of oil tests are lower than those obtained with no oil tests : up to 50 % of surfactant initial concentration (C/Co = 0.4-0.5). But, in every case where a 10 g/l salt preflush was used, a decrease of surfactant retention was observed. A possible explanation could be postulated : an effect is superimposed to that previously described (precipitation of calcic surfactant Ca T_2), and connected with the presence of residual oil, i.e. the partition of surfactant between oil and water. In fact, according to several authors (Sardin and Couliou, 1984 ; Bourdarot and Couliou, 1984), this neutral complex (Ca T_2) has a high oil solubility and a poor water solubility :

$$Ca\ T_2 \rightleftharpoons Ca^{2+} + 2T^- \text{ (dissociation)}$$
$$Ca\ T_2 \rightleftharpoons (Ca\ T_2)$$
$$(oil) \qquad (water)$$

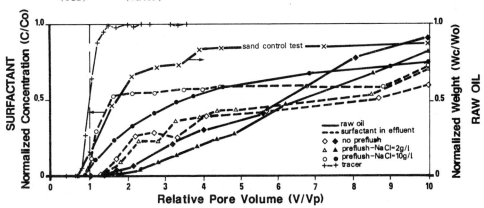

Fig. 7. Influence of the salt preflush on the enhanced gas-oil recovery

So, the amount of exchangeable calcium will be accountable for surfactant partition and that will depend on the preflush technique used. Following a 10 g/l salt preflush, the formation of Ca T_2 complex was limited and the sodic surfactant was in the aqueous effluent. This step corresponded to a high oil recovery. Over two Vp surfactant solution injections,there was still formation of Ca T_2 complex soluble in oil : the salt concentration of injected surfactant solution being 2 g/l (NaCl), the sodium ions concentration was not sufficient to avoid the cation exchange.

CONCLUSIONS

The effects of the residual air and the role of the matrix on the migration of soluble hydrocarbons in aquifer were investigated. Our observations indicate that the propagation of soluble alkanes is dependent on the considered zone. Their retardation is more pronounced in the unsaturated zone than in that saturated, because liquid/gas exchange is the major factor of retardation. Although, the effect of adsorbing character of the matrix exits, this one is less important. On the contrary, the residual air content in soil has no effect on the mobility of soluble aromatic hydrocarbons but the liquid/solid exchange is, in this case, preponderant. In order to minimize this long-term contamination by residual hydrocarbons trapped or adsorbed in soil, our study shows that use of surfactant, as clean·up process, is an effective method providing to control the adsorption or the precipitation of these products. In addition, it indicates that use of a salt preflush step prior to the application of anionic surfactants improves their effectiveness, thus the mobility of non-soluble residual hydrocarbons and their removal yields. This work will be followed in the future by field experiments.

ACKNOWLEDGEMENTS : The authors wish to express their appreciation to the Compagnie Rhénane de Raffinage de Reichstett-Strasbourg and to the Laboratoire du Département de Chimie de l'IUT-Université Strasbourg III (Professors P. Rimmelin and M.A. Hazemann) for their help about analysis and dosage of dissolved hydrocarbons samples.

REFERENCES

Balzer, D. (1988). Oil field injection well stimulation with surfactants. 2ème Congrès Mondial des agents de surface. Paris, pp. 171-182.

Bazin, B., Defives, D. (1984). Détermination de l'adsorption d'un alkylbenzène sulfonate de sodium sur du kaolin par des méthodes statique et dynamique. In: Solid-liquid interactions in porous media, Colloque-bilan, NANCY. Editions TECHNIP, PARIS, pp. 537-556.

Béraud, J.F., Ducreux, J., Gatellier, C. (1989). Use of soil-aquifer treatment in oil pollution control of underground waters. Proceedings of the 1989 Oil Spill Conference, pp. 53-59, API, Washington D.C.

Bourdarot, G., Sardin, M. (1984). Châteaurenard : échange d'ions, mécanismes de la récupération d'huile, rétention des tensio-actifs. In: Solid-liquid interactions in porous media, Colloque-bilan, NANCY. Editions TECHNIP, PARIS, pp. 567-585.

Glover, C.J., Puerto, M.C., Maerker, J.M., Sanvik, E.L. (1978). S.P.E., 7053 of A.I.M.E. Meeting, Tulsa

Hill, H.J. and Lake, L.W. (1977). S.P.E., 6770 of A.I.M.E. Meeting, Denver.

Johnson, R.L., Palmer, C.D., Keely, J.F. (1987). Mass transfer of organics between soil, water and vapor phases : Implications for monitoring, biodegradation and remediation. In: Proceedings of the NWWA/API, Conference on petroleum hydrocarbons and organic chemicals in groundwater. Prevention, detection and restoration, Houston, pp. 493-507.

Rasolofoniaina, J.D., Muntzer, P., Razakarisoa, O. and Zilliox, L. (1988). Impact de l'air résiduel sur le transfert d'hydrocarbures dissous dans l'eau à travers un sable naturel de quartz. Stygologia, 4 (3), 209-227.

Razakarisoa, O., Rasolofoniaina, J.D., Muntzer, P. and Zilliox, L. (1989). Selective dissolution and transport of hydrocarbons in an alluvial aquifer. Role and impact of residual air on groundwater contamination. In: Proceedings of the International Symposium of Contaminant Transport in Groundwater. Stuttgart. A.A. Balkema Publisher/Rotterdam, 405-412.

Sardin, M., Couliou, C. (1984). Rôle des cations sodium et calcium dans le partage d'un tensio-actif anionique entre une huile de gisement stationnaire et de l'eau en écoulement dans un sable argilo-calcaire. In: Solid-liquid interactions in porous media, Colloque-bilan, NANCY. Editions TECHNIP, PARIS, pp. 557-566.

Schwarzenbach, R.P. and Westall, J. (1981). Transport of non polar organic compounds from surface water to groundwater. Laboratory sorption studies. Env. Sci. Tech., 15, 1360-1367.

Schweich, D. (1984). Echange d'ions calcium-sodium, dissolution de la calcite et précipitation d'un tensio-actif anionique dans un sable argilo-calcaire. Expériences dynamiques en colonne. In: Solid-liquid interactions in porous media, Colloque-bilan, NANCY. Editions TECHNIP, PARIS, pp. 63-72.

Texas Research Institute (1979). Underground movement of gasoline on groundwater and enhanced recovery by surfactants. Final report to the API.

Texas Research Institute (1985). Test results of surfactant enhanced gasoline recovery in a large-scale model aquifer. API Publication 4390.

Weast, R.C. (1978). Handbook of Chemistry and Physics. CRC Press, INC. 58th edition.

Zundel, J.P. Siffert, B. (1984). Mécanisme de rétention de l'octylbenzène sulfonate de sodium sur les minéraux argileux. In: Solid-liquid interactions in porous media, Colloque-bilan, NANCY. Editions TECHNIP, PARIS, pp. 447-462.

LONG TERM FATE AND TRANSPORT OF IMMISCIBLE AVIATION GASOLINE IN THE SUBSURFACE ENVIRONMENT

D. W. Ostendorf

Civil Engineering Department, University of Massachusetts, Amherst, MA 01003, USA

ABSTRACT

We measure and model the concentration of separate phase hydrocarbons downgradient of a 19 year old aviation gasoline spill at the US Coast Guard Air Station in Traverse City, Michigan. The separate phase aviation gasoline is presumed to exist in mobile and residual partitions whose transport is modeled as a simple one dimensional balance of storage, advection, volatilization, and linear sorption. Field calibration suggests a retardation factor of 13.1, while volatilization accounts for about 30% of the originally spilled product, underscoring the importance of these two mechanisms in immiscible gasoline fate and transport studies.

KEYWORDS

Nonaqueous phase liquids, VOC contamination, groundwater pollution.

INTRODUCTION

We measure and model the concentration of separate phase hydrocarbons downgradient of a 19 year old aviation gasoline spill at the US Coast Guard Air Station in Traverse City, Michigan. The transport of an immiscible light hydrocarbon spill through the subsurface in its full complexity is an unsteady, nonuniform interaction of liquid product, contaminated water, air, and vapor phases of both liquids, subject to capillarity, differential density effects, and biological and abiological transformation (Schwille, 1967; Pinder and Abriola, 1986). Various components of the overall phenomenon are studied in models with different levels of complexity and mathematical detail. Holzer (1976) describes the unsteady lateral spread of immiscible light oil over an uncontaminated aquifer using an analytical fresh water lens solution, while Osborne and Sykes (1986) include capillary effects in a finite element analysis. Abriola and Pinder (1985) and Baehr and Corapcioglu (1987) propose sophisticated one dimensional finite difference codes of more complete versions of the transport problem, including vapor phases and, in the latter study, biodegradation of an immobile gasoline constituent. Additional analyses could be cited as well.

The simple modeling efforts in this paper are consistent with the spatial and analytical resolution of the available field data at a reasonably well documented aviation gasoline spill site. The study is part of a wider series of US Environmental Protection Agency investigations of dissolved (Rifai et al., 1988) and gaseous (Kampbell et al., 1989) contamination at Traverse City, including source term configuration (Ostendorf et al., 1989), chemical sorption (Bouchard et al., 1989), remedial design (Armstrong and Sammons, 1986), and biodegradation potential (Wilson et al., 1986). Field (Kampbell et al., 1989) and laboratory (Vandegrift and Kampbell, 1988) methods for sampling and analysis have emerged from the Michigan project as well, and major in situ experiments are underway to demonstrate oxygen and nitrate based degradation of the aromatic fractions of the spill.

GOVERNING TRANSPORT EQUATIONS

Following Ostendorf et al. (1989), we consider the separate phase product to exist in mobile M and residual S partitions defined by

$$M = \frac{\text{mass mobile hydrocarbons}}{\text{void volume}} \tag{1a}$$

$$S = \frac{\text{mass residual hydrocarbons}}{\text{mass wet soil}} \tag{1b}$$

$$S = K_D M \tag{1c}$$

where the distribution coefficient K_D is analogous to the parameter for linear sorption. The mobile fraction is presumed to travel as an emulsion with the groundwater flow field near the free surface, while the sorbed partition remains bound to the soil in the contaminated capillary fringe and is in contact with both water and air.

The conservation of hydrocarbon mass (in both partitions) is taken as a balance of storage, advection, and volatilization

$$\frac{\partial}{\partial t}\left(M + \frac{\rho_B S}{n}\right) + \frac{u}{n}\frac{\partial M}{\partial x} + \frac{J}{nh} = 0 \tag{2}$$

with time t, moist bulk density ρ_B, porosity n, contaminated thickness h, evaporation rate J, and specific discharge u in the downgradient x direction. Eqs. 1 and 2 may be combined and expressed in method of characteristics form, so that the depth integrated mobile hydrocarbon concentration in a moving frame of reference is governed by

$$\frac{dM}{dt} = -\frac{J}{R_D nh} \tag{3a}$$

$$\frac{dx}{dt} = \frac{u}{R_D n} \tag{3b}$$

$$R_D = 1 + \frac{K_D \rho_B}{n} \tag{3c}$$

with retardation factor R_D. The frames of reference start (at time t_S) out of a source plane located at the downgradient edge of the spill at x=0 with a known source concentration M_S

$$M = M_S \qquad (t=t_S). \tag{4}$$

The transport problem is thus reduced to separate determinations of the frame concentration and position in the subsurface environment with estimation of the source concentration as a third element of this simple model approach.

Eqs. 3 and 4 yield a straightforward solution when the volatilization rate and specific discharge are assumed constant

$$M = M_S - \frac{J(t - t_S)}{R_D nh} \tag{5a}$$

$$x = \frac{u(t - t_S)}{R_D n}. \tag{5b}$$

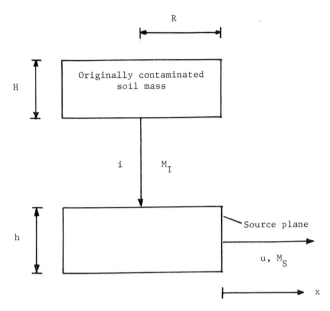

Fig. 1. Source term schematization.

A source term model describing M_S as a function of the frame departure time t_S will complete our analysis.

SOURCE TERM

Physically speaking, the source term may be regarded as mobilized efflux from the soil mass originally contaminated by the aviation gasoline in 1969 at time $t_S=0$. Ostendorf et al. (1989) characterize the source as a disc (radius R and height H) of contaminated soil subject to vertical flushing by episodic exposure to the fluctuating water table, as sketched in Fig. 1. The concentration M_I of mobilized hydrocarbons in the original soil mass is a balance of storage, vertical flushing, and linear sorption

$$nHR_D \frac{dM_I}{dt_S} + iM_I = 0 \tag{6a}$$

$$M_I = M_0 \qquad (t_S = 0) \tag{6b}$$

$$i = \frac{(\eta_{MAX} - \eta_0)n}{1 \text{ year}} \tag{6c}$$

with initial mobilized hydrocarbon concentration M_0. We recognize the periodic nature of water table flushing by basing i on the excursion amplitude of the free surface from its elevation η_0 at the time of the accident to a yearly maximum value η_{MAX}.

Eq. 6 may be integrated from the accident to any subsequent source time with the result

$$M_I = M_0 \exp\left(-\frac{it_S}{R_D nH}\right). \tag{7}$$

We base the constant flushing rate implied by Eq. 7 on the average annual η_{MAX} for the duration of the spill, relying upon available historical data. As suggested by Fig. 1, the contaminated recharge iM_I is input to the flowing groundwater and routed through the source plane 2R by h in size, resulting in an expression for M_S

$$M_S = \frac{\pi R i M_0}{2uh} \exp(-\frac{it_S}{R_D nH}) \tag{8}$$

with M_I specified by Eq. 7.

The solid core data used to calibrate and test the foregoing separate phase transport model is commonly reported as total hydrocarbon concentration T

$$T = \frac{\text{mass total hydrocarbons}}{\text{wet soil mass}} \tag{9a}$$

$$T = \frac{K_D M}{1 - \frac{1}{R_D}} \cdot \tag{9b}$$

Eq. 9b indicates a constant proportionality between mobile and total separate phase hydrocarbons, so that Eqs. 5a and 8 may be expressed in terms of the total mass with the result

$$T = T_S - \frac{JK_D(t - t_S)}{nh(R_D - 1)} \tag{10a}$$

$$T_S = \frac{\pi R i T_0}{2uh} \exp(-\frac{it_S}{R_D nH}). \tag{10b}$$

MODEL CALIBRATION AND TESTING

We calibrate and test the foregoing theory against the solid core data at an aviation gasoline spill site in Traverse City, Michigan. The 1969 release of product presently exists as a separate phase slick occupying a volume about 0.3 m thick, 280 m long, and 80 m wide, with attendant gaseous (Kampbell et al., 1989) and dissolved (Rifai et al., 1988) plumes extending beyond the bounds of immiscible liquid. The contamination rests in uniform fine sand of porosity, permeability, and median grain size of 0.367, 4.7×10^{-11} m^2, and 3.8×10^{-4} m, respectively. The water table slope of 0.004 (Twenter et al., 1985) yields a specific discharge of 1.73×10^{-6} m/s, while an analysis of hydrocarbon vapor pressure profiles in the unsaturated zone implies a volatilization rate of 7×10^{-7} kg/m^2-s.

The historical water table fluctuations are inferred from water surface elevations of Lake Michigan due to the proximity of the site to Grand Traverse Bay, about 1300 m to the northeast. Twenter et al. (1985) establish a close correlation between lake levels and site water table elevations in this regard. Since the analysis rests on relative fluctuations, we adopt observations maintained by the National Oceanic and Atmospheric Administration at Ludington, Michigan some 60 miles south of the site area along the eastern shore of Lake Michigan. The data indicate a December 1969 average elevation n_0 of 578.91 ft (International Great Lakes Datum) and the maximum monthly average for each year is used to estimate a representative n_{MAX} value of 580.24 ft. Eq. 6c then suggests that i is 4.71×10^{-9} m/s.

We calibrate our source term concentrations with solid core data reported by Ostendorf et al. (1989), who observe that the 1988 total hydrocarbon concentration of the originally contaminated soil mass is 0.00520 (kg product/kg wet soil). This value is set equal to T_I at source time $t_S = 5.99 \times 10^8$ s. A calibrated T_0 value of 0.0354 follows from Eq. 10b, with H=0.306 m. Ostendorf et al. (1989) also calculate a retardation factor of 13.1, based on source flushing and slow transposition of the upgradient edge of contamination. This value is confirmed by the present investigation: Eq. 5b suggests that the immiscible plume extends 216 m downgradient of the source plane in 1988, consistent with the data at stations 50R and 50AK, as cited in Table 1. Eq. 3c yields a distribution coefficient of 0.00222 m^3/kg, based on a wet bulk density of 2000 kg/m^3.

Solid core samples are obtained from the Traverse City site at the 8 stations sketched in Fig. 2 using the US Environmental Protection Agency coring protocol put forth by Leach et al. (1988). A 0.0509 m diameter stainless steel core barrel equipped with a wireline vacuum seal is driven ahead of hollow stem augers to obtain a relatively undisturbed sample, which is extruded into Mason jars inside a nitrogen filled glove box. We augment the sampling technique by using a

TABLE 1 Predicted and Observed Total Hydrocarbon Concentrations

Station	t $sx10^8$	x m	t_S $sx10^8$	T_S	T(pred.)	T(meas.)	δ %
50AN-25	6.07	40	4.96	0.0041	0.0028	0.0011	-41
50AI-7,15	5.99	80	3.77	0.0061	0.0035	0.0045	16
50AB-9,10	5.94	131	2.30	0.0098	0.0055	0.0029	-27
50AM-15-17	5.99	157	1.63	0.0120	0.0069	0.0155	72
50AG-15,16	5.99	170	1.27	0.0135	0.0080	0.0075	-4
50AS-29	6.07	195	0.65	0.0164	0.0100	0.0200	61
50R-6,7	5.81	207	0.06	0.0198	0.0131	0.0087	-22
50AK	5.99	267	----	------	------	BDL*	--

*Below detection limit.

hydrocarbon sniffer to analyze the minimal jar headspace within the glove box--the highest VOC levels correspond to the interval of maximum residual contamination in each of the stations tested. The samples are packed in ice, returned to the Robert S. Kerr Environmental Research Laboratory, stored at 4 deg C, and analyzed in an FID wide bore capillary column gas chromatograph after methylene chloride extraction, following the methodology of Vandegrift and Kampbell (1988).

The data appear in Table 1, which summarizes our test results. The core sections with the maximum total hydrocarbon concentration are reported in the table; in nearly all cases these are bound by soil with much lower levels of contamination. We assess model accuracy using the mean $\bar{\delta}$ and standard deviation σ of the error δ defined by (Benjamin and Cornell, 1970)

$$\delta = \frac{T(measured) - T(predicted)}{T_S} \tag{11a}$$

$$\bar{\delta} = \frac{1}{j}(\Sigma\delta) \tag{11b}$$

$$\sigma = [\frac{1}{j}(\Sigma\delta^2) - \bar{\delta}^2]^{1/2}. \tag{11c}$$

The 8% mean error implies a slight underprediction of the data base, while the 41% standard deviation reflects a modest level of test accuracy, indicative of the scatter inherent in the sampling and analysis of a 0.3 m contaminated interval located some 4 m below the ground surface. The results are judged to be quite satisfactory in view of the simplicity of the model approach.

DISCUSSION

The calibrated source concentration leads to a calculation of the mass m_O of aviation gasoline originally spilled into the subsurface environment

$$m_O = \pi\rho_B R^2 H T_O \tag{12a}$$

$$m_O = 1.09 \times 10^5 \text{ kg}. \tag{12b}$$

Eq. 12b is somewhat higher than the estimate of Ostendorf et al. (1989), whose 64,000 kg figure rests on a literature value of 0.212 for T_O. The revision reflects additional downgradient solid core data and a separate phase transport model.

An appreciable fraction of this contamination evaporates over the 19 year life of the plume. The volatilized mass m_V may be readily computed from the following equation

$$m_V = 2R \int_0^t JL \, dt' \tag{13a}$$

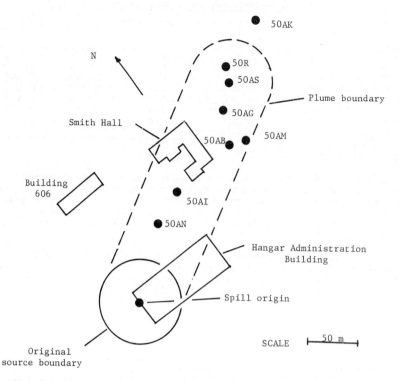

Fig. 2. Solid core sampling stations at Traverse City, Michigan.

$$L = \frac{ut}{R_D n} \tag{13b}$$

with plume length L. Eq. 13 may be simply integrated for a presumedly constant volatilization rate with the result

$$m_V = \frac{JRut^2}{R_D n} \tag{14a}$$

$$m_V = 3.62 \times 10^4 \text{ kg} \qquad (t = 5.99 \times 10^8 \text{ s}). \tag{14b}$$

Thus, about 33% of the aviation gasoline has evaporated in the 19 year life of the plume, and 7.24×10^4 kg of product presently (December 1988) occupies the contaminated capillary fringe. Eq. 14a indicates about 10^9 s (33 years) will be required for total volatilization of the gasoline, under the assumption of a constant volatilization rate and a traveling front of pollution. The separate phase plume would extend about 370 m from its source at this time.

The significance of volatilization to the transport of aviation gasoline contrasts markedly with the negligible importance of dissolution, as has been noted by Baehr (1987). We assess the relative importance of the two fluxes by evaluating the following ratio

$$\frac{iC_s}{J} < 10^{-2} \tag{15}$$

with solubility C_s of aviation gasoline taken as 0.004 kg/m^3, based on the known composition at the site (Ostendorf et al., 1989). It should be remarked that the two transport mechanisms may be of comparable importance for the soluble (BTX) fraction of the product, however. These compounds comprise about 6% of the spill at Traverse City.

CONCLUSIONS

We measure and model the concentration of separate phase hydrocarbons downgradient of a 19 year old spill of 1.09×10^5 kg of aviation gasoline at the US Coast Guard Air Station in Traverse City, Michigan. The separate phase gasoline is presumed to exist in mobile and residual partitions whose transport is modeled as a simple one dimensional balance of storage, advection, volatilization, and linear sorption. Data from the previous source characterization study of Ostendorf et al. (1989) are used to calibrate values for the retardation factor (R_D= 13.1) and initial source concentration (T_o=0.0354), while vadose zone hydrocarbon vapor pressure profiles yield an estimate of the volatilization rate ($J=7.0 \times 10^{-9}$ kg/m^2-s). Downgradient solid core data test the calibrated model with a modest degree of accuracy, reflected in error mean and standard deviation values of 8% and 41%, respectively. The strong retardation and appreciable (30%) amount of evaporation underscore the importance of sorption and volatilization in immiscible gasoline fate and transport studies.

ACKNOWLEDGEMENT

The National Center for Groundwater Research of Rice University provided sabbatical support for Dr. Ostendorf, who served as a Visiting Associate Professor at the Robert S. Kerr Environmental Research Laboratory of the US Environmental Protection Agency from September 1988 to June 1989. The solid core analyses were performed by NSI Technology Services Inc., and we wish to acknowledge field support by the US Coast Guard. Although the study was conducted at Kerr Laboratory, the paper has not been subjected to USEPA review. Accordingly, it does not necessarily reflect the views of the Agency and no official endorsement should be inferred.

REFERENCES

Abriola, L.M. and Pinder, G.F. (1985). A multiphase approach to the modeling of porous media contamination by organic compounds. Water Resources Research, 21, 11-18.
Armstrong, J.M. and Sammons, J.H. (1986). Assessment and management of a 15 year old VOC groundwater contaminant plume. Proceedings Petroleum Hydrocarbons and Organic Chemicals in Groundwater, NWWA/API, Houston, TX, pp. 797-811.
Baehr, A.L. (1987). Selective transport of hydrocarbons in the unsaturated zone due to aqueous and vapor phase partitioning. Water Resources Research, 23, 1926-1938.
Baehr, A.L. and Corapcioglu, M.Y. (1987). A compositional multiphase model for groundwater contamination by petroleum products 2--Numerical solution. Water Resources Research, 23, 201-213.
Benjamin, J.R. and Cornell, C.A. (1970). Probability, Statistics, and Decision for Civil Engineers. McGraw-Hill, New York, NY, 684 pp.
Bouchard, D.C., Enfield, C.G., and Piwoni, M.D. (1989). Transport processes involving organic chemicals. Reactions and Movement of Organic Chemicals in Soils, Soil Science Society of America, Madison, WI, pp. 349-371.
Holzer, T.L. (1976). Application of a groundwater flow theory to a subsurface oil spill. Groundwater, 14, 138-145.
Kampbell, D.H., Wilson, J.T., and Ostendorf, D.W. (1989). Simplified soil gas sensing techniques for plume mapping, remediation monitoring, and degradation modeling. Proceedings Fourth National Conference on Petroleum Contaminated Soils, Lewis Publishers, Chelsea, MI, in press.
Leach, L.E., Beck, F.P., Wilson, J.T., and Kampbell, D.H. (1988). Aseptic subsurface sampling techniques for hollow stem auger drilling. Proceedings Second National Outdoor Action Conference on Aquifer Restoration, NWWA, Las Vegas, NV, pp. 31-51.
Osborne, M. and Sykes, J. (1986). Numerical modeling of immiscible organic transport in the Hyde Park Landfill. Water Resources Research, 22, 25-33.
Ostendorf, D.W., Kampbell, D.H., Wilson, J.T., and Sammons, J.H. (1989). Mobilization of aviation gasoline from a residual source. Research Journal Water Pollution Control Federation, in press.
Pinder, G.F. and Abriola, L.M. (1986). On the simulation of nonaqueous phase organic compounds in the subsurface. Water Resources Research, 22, 109S-119S.
Rifai, H.S., Bedient, P.B., Wilson, J.T., Miller, K.M., and Armstrong, J.M. (1988). Biodegradation modeling at aviation fuel site. Journal of Environmental Engineering, 114, 1007-1029.

Schwille, F. (1967). Petroleum contamination of the subsoil--a hydrological problem. *The Joint Problems of the Oil and Water Industries*, Institute of Petroleum, New York, NY, pp. 23-54.

Twenter, F.R., Cummings, T.R., and Grannemann, N.G. (1985). Groundwater contamination in East Bay Township, Michigan. *WRIR 85-4064*, USGS, Lansing, MI, 63 pp.

Vandegrift, S.A. and Kampbell, D.H. (1988). Gas chromatographic determination of aviation gasoline and JP-4 jet fuel in subsurface core samples. *Journal Chromatographic Science*, **26**, 566-569.

Wilson, B.H., Bledsoe, B.E., Kampbell, D.H., Wilson, J.T., Armstrong, J.M., and Sammons, J.H. (1986). Biological fate of hydrocarbons at an aviation gasoline spill site. *Proceedings Petroleum Hydrocarbons and Organic Chemicals in Groundwater*. NWWA/API, Houston, TX, pp. 78-90.

FATE AND TRANSPORT OF PETROLEUM IN THE UNSATURATED SOIL ZONE UNDER BIOTIC AND ABIOTIC CONDITIONS

J. B. Carberry and S. H. Lee

Department of Civil Engineering, University of Delaware, Newark, DE 19716, USA

ABSTRACT

Two different types of soil were selected to observe the migration and/or degradation of petroleum contaminants. One soil sample was composed of fine clay and was contaminated by an electrical insulating oil. The other soil sample was a coarse soil contaminated by fuel oil number 2 and number 6. A portion of each was cultured in a minimal medium to generate large concentrations of indigenous microbes capable of degrading each petroleum contaminant. Four columns fabricated of plexiglass with five ports at 6 inch intervals were filled with the contaminated soil which had been air dried and autoclaved for sterilization. These columns simulated unsaturated soil in the vadose zone.

Two types of experiments were carried out during this research. In one type of experiment, the petroleum was added just once in order to simulate a petroleum spill. In the other type of experiment, petroleum was added to the column repeatedly at measured intervals in order to simulate a petroleum leak. To one of the two columns operated for each soil, the contaminating petroleum and water were added. To the second column operated for each soil, the contaminating petroleum, water, and cultured microbes were added. From these experiments, the biodegradability was observed to be affected by soil composition and conditions, concentrations of petroleum contaminants, and microbial concentrations. Better microbial degradation occurred in the fine soil for the simulated spill case than in the coarse soil. In this soil with low porosity, the microbes were confined to the top of the column and the petroleum concentration was low. In the simulated leak case, the biodegradation took place at a higher rate in the coarse soil than in the fine soil. Here, the high rate of advection distributed the microbes throughout the depth of the column and had a diluting effect on the high petroleum cumulative concentrations.

From these experiments, a mass balance of petroleum contamination could be determined at all time intervals at all column depths. The sterile column behavior exhibited only physical changes such as adsorption and evaporation and could be compared to results in columns of biotic conditions. These differences permitted the development of a computer model to simulate and predict physical and biological processes in fine and coarse soils.

INTRODUCTION AND BACKGROUND

Laboratory scale columns containing isotropic samples of two petroleum-contaminated soils were operated in order to simulate effects of a petroleum leak and effects of a petroleum spill. Five ports throughout the total depth of the columns were used to periodically measure volatilized vapor phase petroleum concentrations and to remove suspension containing liquid and solid phase petroleum concentrations. Columns were operated under sterile conditions and under conditions in which selected microbial communities were added. Three phase petroleum measurements with time and depth were obtained in order to quantify threats to groundwater described by Cohen (1986), Grenny, et al. (1987), Symons, et al. (1988) and Short (1988).

Mechanisms controlling the fate of such hydrocarbon leaks and spills have been described by Short (1988) and Letey and Oddson (1972). Petroleum hydrocarbons and their related fates have been studied by Mackey (1985), Landrum et al. (1987), Vandermenleu (1987) and others. From these previous studies and from this study, a one dimensional computer model was developed from petroleum mass balance measurements in order to predict the magnitude of adsorption/desorption, solubility and advection, volatilization and biodegradation. The purpose of this study was to indicate the overall fate and transport of petroleum due to the variability of these mechanisms under differing conditions.

METHODOLOGY

Laboratory Experiments

Preliminary experiments were carried out to select indigenous microbes from two contaminated soil samples from nearby electric power plants. One fine clay soil was contaminated with mineral lubricating oil and the other coarse soil was contaminated with Numbers 2 and 6 fuel oil. In order to select indigenous microbes from these two soil samples, two stock cultures were developed by suspending 10 grams of contaminated soil in 20 ml of autoclaved minimal salts medium in sterile 125 ml erlenmeyer flasks fitted with screw caps. The contaminating petroleum was added to each respective flask to serve as the sole carbon source. These flasks were shaken intermittently and fed with petroleum periodically for one month. At the end of this period, the soil was settled out and the suspended microbes decanted into a flask of 200 ml of sterile minimal salts medium. Each culture was periodically shaken and fed with its respective petroleum contaminant until stock cultures containing a highly-concentrated microbial community for each contaminated soil were obtained for column experiments.

Preliminary soil testing and preparation were carried out in order to characterize each soil and to provide appropriate soil for the column experiments. Each contaminated soil was tested for nitrogen, phosphate, potash and pH using the soil test kit by Luster leaf Manual for Soil Test (1988). Particle size distribution, moisture content, specific gravity, porosity, specific area and total surface area per control volume were measured according to ASTM (1985) and Das (1986).

Each contaminated soil was prepared for column use by washing, air drying, gentle grinding and autoclaving repeatedly. This technique was used in order to sterilize the soil in a benign manner to preserve each individual soil character. Following the final treatment cycle, the soil was wetted and packed in columns of 36 in. length and 1 and 3/4 in. diameter. Each port was fitted with 5 ports spaced approximately every 6 inches in vertical column depth. Four columns were used per experiment, 2 per contaminated soil. One column for each soil was kept abiotic, using the prepared sterile soil. The second column was operated under biotic conditions in which the previously-described selected microbial community was added periodically to each soil. Each column was fitted with a removable-mesh screen at the bottom to facilitate packing and unloading the columns. Any liquid passing through the screen during an experiment was collected below the columns.

Simulated petroleum spill experiments were run by adding to the top of the abiotic column of each soil 50 ml of water and 1.0 ml of petroleum at time zero. Every three days 50 ml of water was added to the top of each abiotic column. The biotic column of each soil samples operated in the same manner with the addition of 2 ml of concentrated microbes selected from each respective soil/type, as described above, with each addition of water. Petroleum leak experiments were operated similarly except that for this simulation, 0.5 ml petroleum aliquots were added every three days with the 50 ml water addition.

Sampling was carried out every three days for all experiments which lasted about 22 days. First, volatile petroleum gas phase measurements were carried out at each port by a TIP-II with a photoionization detector. Next each column was tipped to a 45° angle and each port was opened sequentially to permit interstitial liquid and some soil to flow into receiving test tubes. One ml from each sample of removed suspension was then serially diluted for plate counting on trypticase soy agar (TSA). Ten ml of remaining suspension was filtered through 0.22 μm pore size tared millipore filter paper (as the selected microbes were very small and penetrated conventional 0.45 μm pore size filters). The filter papers and solids were dried at 105°C for 2 hours, cooled and re-weighed to determine the solids concentration by weight. The dried solids and paper were then combusted in tared covered crucibles at 600°C, cooled, and re-weighed to determine the organic content. The total microbial weight was calculated from the microbial plate count above and subtracted from the gravimetric determination of organic content in order to determine the weight of adsorbed petroleum on the soil particles. The filtrate was analyzed for soluble COD by the HACH Microdigestion Method (Gibbs, 1982), and for DO and pH using an Orion 97-08 model oxygen electrode and Corning model 476541 pH electrode with a Fisher model 825 MP Accumet meter.

Computer Modeling Approach

All the data were logged into an IBM Enhanced AT computer and analyzed from Lotus II spread sheets. Petroleum concentrations in all phases were regressed linearly and logarithmically with time to determine the best fit for all biotic and abiotic conditions, for both the fine clay soil and the coarse soil, at all depths in the columns for both the petroleum leak and petroleum spill experiments. The slopes of these plots with time or depth yielded constants which were then fitted to terms in the following transport model equation for adsorption, advection, volatilization and biodegradation:

$$\frac{\partial C}{\partial t}\partial Z = K_t C_L V_\phi dZ + K_S C_S A_S \rho dZ + K_G C_V dZ + \mu_B C_L dZ \qquad (1)$$

where K_t = advective rate constant in liquid phase (day^{-1})
V_ϕ = pore space (cm^3 void/cm^3 soil)
K_S = mass adsorption/desorption rate constant (day^{-1})
C_S = mass adsorbed petroleum per mass solid (mg/mg)
A_S = specific surface area (cm^2/cm^3)
ρ = bulk density (gm/cm^3)
K_G = volatilization rate constant (day^{-1})
μ_B = biodegradation rate constant (day^{-1})
C_L and C_V = concentration of petroleum in the liquid and vapor phases, respectively

Predicted values were then compared with measured values in all phases.

RESULTS AND DISCUSSION

Soil Characterization

The soils are characterized in Table 1.

Table 1

		Soil 1	Soil 2
Sieve Analysis	%>2mm	9.6	43.48
	2<%<0.2	67.42	47.87
	%<0.2	22.29	7.96
Physical Characteristics	%Moisture	18.25	8.06
	Specific Gravity	2.77	2.62
	Porosity	0.48	0.53
	Specific Surface Area	83.5 cm^2/gm	43.15 cm^2/gm
	Surface Area per Control Volume	16.3 m^2	8.0 m^2
Chemical Characteristics	pH	5.5	6.5
	Nitrogen	M	M
	Phosphate	H	H

Soil 1 was composed mainly of fine clay which retained moisture, had a large surface area and a rather low pH. Soil 2 was composed of coarse particles and contained approximately half the moisture content and surface area as Soil 1. Soil 2 had a higher pH than Soil 1, and both soils contained adequate nitrogen and phosphate to support microbial biodegradation activity.

Column Experiments

The petroleum leak experiments indicated that advective transport was a controlling mechanism for decreasing liquid phase petroleum concentration in the fine soil, while both advection and biodegradation were significant in the coarse soil. See Figure 1. Little microbial growth occurred in the fine soil because high petroleum concentrations prevented microbial growth until late in the experiment when microbes became acclimated to high concentrations. See Figure 2. The coarse soil supported bacterial growth at all depths during the entire experimental period indicating that microbes could penetrate the entire column depth through larger pore spaces. A relatively constant pH was maintained in both soils throughout the experimental period, and the DO was maintained between 3 and 7 mg/l during all phases of the experiments at all depths. In both the fine and coarse soils, the concentration of adsorbed petroleum was less in the columns receiving selected microbial additions than in the abiotic columns. See Figure 3. The gas phase concentrations were very different for the fine soil and coarse soil. Large concentrations of petroleum vapor built up in both the biotic and abiotic columns of the coarse soil and then dissipated after 10 days. In the fine soil, large vapor phase concentrations built up toward the end of the experiments in the abiotic column, values as high as 300 mg/l. In the biotic column, however, concentrations were kept to a minimal value, usually less than 50 mg/l. See Figure 4.

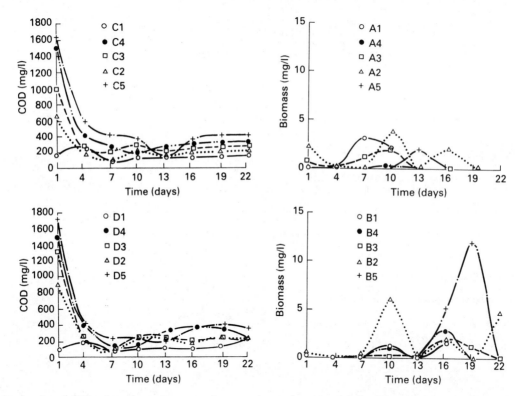

Figure 1. Soluble Petroleum Concentration for Leak Experiment in Coarse Soil, Abiotic Condition (C) and Biotic Condition (D).

Figure 2. Microbial Concentrations in Fine Soil During Leak Experiment, Abiotic Condition (A) and Biotic Condition (B).

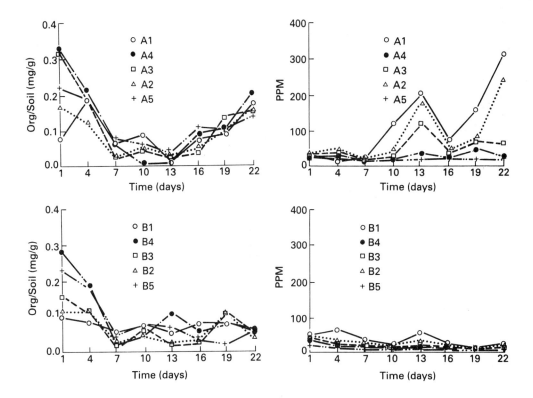

Figure 3. Adsorbed Petroleum Concentration in Fine Soil During Leak Experiment, Abiotic Condition (A), Biotic Condition (B).

Figure 4. Gas Phase Petroleum Concentration for Fine Soil During Leak Experiment, Abiotic Condition (A), Biotic Condition (B).

For the oil spill experiments, significant biodegradation occurred at the top of the fine clay soil column where microbes lodged due to filtration effects. With only one petroleum addition, the concentration was low enough and the concentrations of microbes at the top were high enough that biodegradation was located here at the beginning of the experiment. In the coarse soil, advection was so significant that the petroleum concentration was reduced to a level low enough to reduce the first order rate of biodegradation. Microbial counts corroborate this possibility, for growth occurred to higher microbial concentrations in the fine soil. Again, for these experiments, the pH remained quite constant in the 7-8 region and DO concentration rose from initial concentrations around 4 mg/l to about 8 mg/l at the end of the experiment when all the petroleum had been degraded. Little adsorption occurred on the fine clay soil under abiotic conditions, but there was significant increase in the solid phase petroleum concentrations toward the end of the experiments in both the fine clay and coarse soils. This result was unexpected from the differences obtained in surface area analysis. See Figures 5 and 6. In Figures 5 and 6, the build up of petroleum concentrations beginning at Day 10 in the biotic columns coincided with microbial concentration decreases. The gas phase results for the spill experiments were almost the same as for the previous leak experiments.

Computer Modeling Results

Laboratory data were utilized to determine rate constants for each mechanism described in Equation 1. The model was more successful in predicting results under biotic conditions than under abiotic conditions for both the leak and spill experiments. Correlation coefficients were consistently high for biotic conditions, but for the abiotic conditions in the leak experiments, correlation coefficients were low at the top of the columns of both fine and

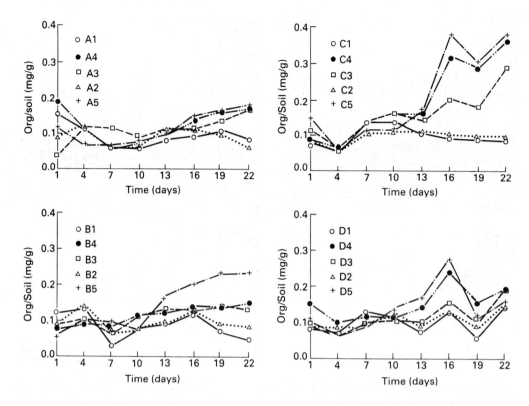

Figure 5. Adsorbed Petroleum Concentration in Fine Soil During Spill Experiment, Abiotic Condition (A), Biotic Condition (B).

Figure 6. Adsorbed Petroleum Concentration in Coarse Soil During Spill Experiment, Abiotic Condition (A), Biotic Condition (B).

coarse soils. In contrast correlation coefficients were low at the bottom depths of the abiotic columns for the spill experiments. These variations in results indicate that porosity differences with depth must influence the model performance. These porosity differences with depth apparently were moderated under biotic conditions when microbial growth diminished all pore spaces. See Figures 7, 8, 9 and 10.

CONCLUSIONS

1. Biodegradation took place in the liquid and solid phases of both fine and coarse soil.
2. Adsorption took place equally well in fine and coarse soils and caused increasing solids concentrations during leak experiments.
3. Advection was a very important transport mechanism in both leak and spill models.
4. Volatization was more easily detected in the fine soil which impeded its escape from the top of the columns.
5. DO and pH values never inhibited respiration activity of the microbes.

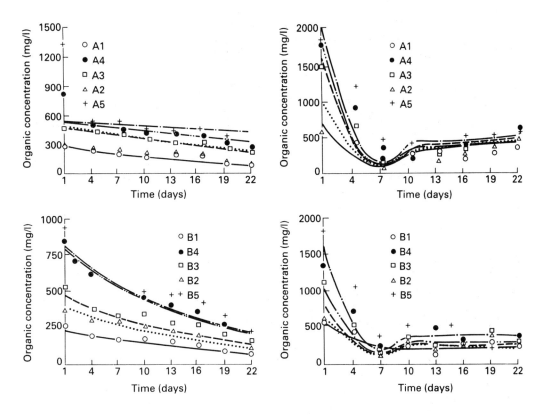

Figure 7. Predicted Petroleum Liquid Phase Concentrations (lines) Developed by Computer Model and Measured Values (points) for Fine Soil.

Figure 8. Predicted Petroleum Liquid Phase Concentrations (lines) Developed by Computer Model and Measured Values (points) for Fine Soil During Leak Experiment.

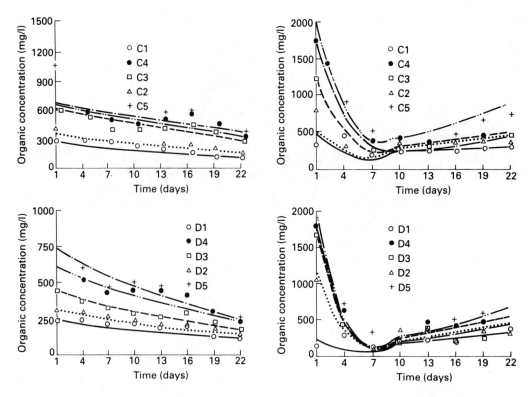

Figure 9. Predicted Petroleum Liquid Phase Concentrations (lines) Developed by Computer Model and Measured Values (points) for Coarse Soil During Spill Experiment, Abiotic (C) and Biotic (D).

Figure 10. Predicted Petroleum Liquid Phase Concentrations (lines) Developed by Computer Model and Measured Values (points) for Coarse Soil During Leak Experiment, Abiotic (C) and Biotic (D).

REFERENCES

American Society for Testing and Materials. 1985d. Standard test method for particle-size analysis of soils. D422-63. 04.08:117-127.

Cohen, Y. (1986). Organic pollutant transport. Environ. Sci. Technol. 20, 538-544.

Das, B.M. (1986). Specific Gravity. In: Soil Mechanics Laboratory Manual, Engineering Press, San Jose, CA, pp 9-13.

Gibbs, C.R. (1982). Introduction to Chemical Oxygen Demand. HACH Technical Center for Applied Analytical Chemistry, 8, 1-19.

Grenny, W.J., Caupp, C.L. and Sims, R.C. (1987). A mathematical model for the fate of hazardous substances in soil: model description and experimental results. Hazardous Waste and Hazardous Materials 4, 223-239.

Landrum, P.F., Giesy, J.P., Oris, J.T. and Allred, P.M. (1987). Photoinduced toxicity of polycyclic aromatic hydrocarbons to aquatic organisms. In: Oil in Freshwater, J.H. Vandermeulen and S.E. Hrudey (Eds.) Pergamon Press, New York, pp 304-318.

Letey, J. and Oddson, J.K. (1972). Mass transfer. In: Organic Chemicals in the Soil Environment, C.A.I. Goring and J.H. Hamaker (Eds.). Marcel Dekker, New York, pp 399-442.

Mackey, D. (1985). The physical and chemical fate of spilled oil. In: Petroleum Effects in the Arctic Environment, F.R. Englehardt (Ed). Elsevier, London and New York. pp 37-62.

Manual for Soil Test. (1988). Luster Leaf Products, Inc. Crystal Lake, IL.

Short, T.E. (1988). Movement of contaminants from oily wastes during land treatment. In: Soils Contaminated by Petroleum. Environmental and Public Health Effects E.J. Calabrese and P.T. Kostecki (Eds). John Wiley & Sons, pp 317-330.

Symons, B.D., Sims, R.C. and Grenney, W.J. (1988). Fate and transport of organics in soil: Model predictions and experimental results. J. Wat. Poll. Cont. Fed. 63 1684-1693.

BIODEGRADATION OF BTEX IN SUBSURFACE MATERIALS CONTAMINATED WITH GASOLINE: GRANGER, INDIANA

J. M. Thomas, V. R. Gordy, S. Fiorenza and C. H. Ward

National Center for Ground Water Research, Department of Environmental Science and Engineering, Rice University, Houston, TX 77251, USA

ABSTRACT

The microbial ecology and potential for biodegradation of benzene, toluene, ethylbenzene, and o- and m-xylene (BTEX) in core materials contaminated with unleaded gasoline were investigated. The site studied was unique because a portion of the contaminated area was biostimulated in a demonstration of the use of hydrogen peroxide as an oxygen source in *in situ* biorestoration. Two years after termination of the field demonstration, core samples were collected from uncontaminated, contaminated, and biostimulated areas at the site and analyzed for inorganic nutrients, microbial numbers, mineralization potential of glucose, benzene, and toluene using liquid scintillation counting, and biotransformation of BTEX using gas chromatography. The results indicated that the subsurface microflora at the site was active and capable of degrading a variety of compounds. Microbial numbers and contaminant biodegradation potential in samples from the biostimulated area were greater than in uncontaminated and contaminated zones. Toluene, ethylbenzene, and m-xylene were removed in all core materials, whereas o-xylene was recalcitrant. Mineralization experiments indicated that toluene was mineralized to a greater extent than benzene. These data indicated that the biodegradation potential of the subsurface material from the biostimulated zone, which still contained residual hydrocarbon, remained enhanced for at least 2 yr after the *in situ* biorestoration process had been terminated.

KEYWORDS

Biodegradation; BTEX; gasoline; ground water; subsurface; biorestoration

INTRODUCTION

Contamination of the subsurface environment with hazardous materials which leak from above ground and underground storage tanks is a growing national concern. As a result, several physical, chemical, and biological methods for subsurface remediation have been developed (Thomas et al., 1987a). The biological methods include above ground and subsurface treatments. Above ground treatment involves withdrawal of the contaminated ground water and treatment at the surface, after which the treated water is disposed or injected into the subsurface with added nutrients and sometimes seed bacteria. Biological subsurface treatment, *in situ* biorestoration, is the process by which the indigenous subsurface microflora is stimulated to degrade the contamination by the addition of inorganic nutrients and a terminal electron acceptor (Thomas and Ward, 1989). Successful application of *in situ* biorestoration requires the presence of contaminant-degrading microorganisms in subsurface materials through which a nutrient solution can be transported.

In 1977, an above ground storage tank leaked unleaded gasoline into the subsurface at a storage facility located in Granger, Indiana. The gasoline percolated into a shallow aquifer located 25 ft below the surface. Remedial action involved recovery of the removable, unaltered phase of gasoline. The site was unique because a portion of the contaminated area was used for an American Petroleum Institute-sponsored pilot-scale demonstration of *in situ* biorestoration in May 1984 (Minugh et al., 1987). The demonstration was designed to assess the use of hydrogen peroxide as an oxygen source in subsurface bioremediation. However, the demonstration was not taken to completion and considerable hydrocarbon remained in the test plot.

The present study investigated the ecology and biodegradation potential of the microflora in subsurface materials collected from the site. Specifically, microbial numbers, the heterotrophic potential, and the biodegradation potential of benzene, toluene, ethylbenzene, o-xylene, and m-xylene (BTEX) were determined in subsurface samples from uncontaminated, contaminated, and biostimulated zones at the site.

MATERIALS AND METHODS

Site Characterization and Sample Collection

The subsurface formation was glacial in origin and consisted of fine-grained unconsolidated outwash sand (Minugh et al., 1987). Information from core samples collected from the biostimulated area identified the texture, grain size, microstratigraphy, and density of subsurface materials as heterogeneous. Sieve analyses using representative core samples indicated that the sand was a silty, medium to fine-grained, well rounded quartz sand with some silt and gravel. Hydrogeological parameters measured in the biostimulated area indicated that the natural gradient of the water table was 0.002 ft/ft with an average ground water velocity of 0.0068 ft/day. Depth to the water table of the unconfined aquifer was 25 ft. The hydraulic conductivity was 4.8 ft/day and the transmissivity was about 120,000 gpd/ft.

Core samples and ground water were collected aseptically from uncontaminated, contaminated, and biostimulated zones (Figure 1) and stored at 5°C until used. Cores were collected using the method developed by Dunlap et al. (1977) and modified by Wilson et al. (1983). Ground water was collected by first clearing wells of 3 to 5 well volumes and then pumping water through sterile tubing into sterile containers.

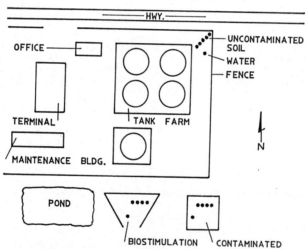

Fig. 1. Terminal area plot plan, Granger, IN.

Core material from uncontaminated, contaminated, and biostimulated zones was sent to a commercial laboratory (Galbraith Laboratories, Knoxville, TN) for analysis of total carbon, carbonate, nitrate, nitrite, and phosphate. The pH of the core materials and temperature of the ground water were determined.

Microbial Numbers

Using the spread plate technique, microbial numbers were determined on Nutrient Agar (Difco Industries, Detroit, MI) and two types of solid media prepared using 1.5% Noble Agar (Difco Industries) and uncontaminated or contaminated ground water collected from the site. The pH of the ground water agars was adjusted to 8.0, the pH of the subsurface material at the site. Cells were released from the core material by shaking 10 g of wet solids in 95 ml of sterile 0.1% sodium pyrophosphate, after which the resulting suspension was diluted serially and then plated (Ghiorse and Balkwill, 1983).

Chemicals

The following ^{14}C radioactive chemicals were purchased from Pathfinders Laboratories, Inc., St Louis, Missouri: benzene-ring-UL-^{14}C, 10.0 mCi/mmol; toluene-ring-UL-^{14}C, 10.9 mCi/mmol; and D-[$^{14}C(U)$] glucose, 1.88 mCi/mmol. All labeled compounds were at least 98% pure and all unlabeled compounds were of the highest purity obtainable.

Biotransformation Studies

The disappearance of toluene, ethylbenzene, m-xylene, and o-xylene was determined in experiments using a modification of the test tube microcosm system developed by Wilson et al. (1983). A slurry was prepared by adding 20 g of core material and 33 ml of a sterile solution amended with 300 μg/L toluene, ethylbenzene, o-xylene, and m-xylene in a methanol carrier to 40 ml screw-cap test tubes. Sterile controls were prepared using autoclaved core material and a sterile solution which had been amended with 1% sodium azide. The microcosms were vortexed to thoroughly mix their contents and triplicates of both nonsterile and sterile microcosms were destructively sampled after desired periods of incubation. The aqueous phase was extracted by passing 33 ml of liquid through a ^{18}C Sep-Pak (Waters Associates, Milford, MA). The remaining solids were extracted by adding 20 ml of organic free water to the microcosm, shaking for 15 min, centrifuging the resultant mixture, and then passing the supernatant fluid through the same Sep-Pak used to extract the aqueous phase. The compounds concentrated in the Sep-Pak were eluted with methylene chloride. At some sampling points, the dissolved oxygen concentration of a portion of the aqueous phase was measured with an oxygen probe and meter (YSI model 57, Yellow Springs, OH). Standards were added to the samples to determine extraction and injection efficiencies. The Tracor gas chromatograph (Tracor, Austin, TX), Model 560 was equipped with a flame ionization detector. The injector and detector were held at 225 and 250°C, respectively, and the flow rate of the carrier gas, helium, was 1 ml/min. Samples were injected onto a 30 m fused silica, crosslinked with FFAP megabore capillary column (Hewlett Packard, Avondale, PA) with a 0.53 mm i.d. The gas chromatograph was programmed to hold initially at 50°C for 5 min, increase at a rate of 10°C/min to 200°C, and then hold for 5 min. A Spectra Physics (San Jose, CA) model 4100 integrator was used to quantify the concentrations of the target compounds.

Mineralization Studies

Microcosms (25 ml glass scintillation vials) were filled to capacity with 10 g of subsurface material and sterile ground water, sealed with Teflon-lined septa and open-top screw caps, and then amended with 2 ng/L ^{14}C-benzene or toluene in 0.5 μL trichloro-ethylene. Sterile controls were prepared using subsurface material which had been autoclaved and sterile ground water to which 1% sodium azide had been added. All samples were run in triplicate. The vials were incubated for designated periods of time which differed for each compound, and then sampled for the production of $^{14}CO_2$. The microcosms were sampled by adding three drops of concentrated sulfuric acid and then purging the vials through a series of traps using a modified method of Marinucci and Bartha (1979) which is described elsewhere (Thomas et al., 1987b). The trapped $^{14}CO_2$ was then counted on a Beckman liquid scintillation counter (Model 3801, Irvine, CA) after the samples were stored in the dark for 1 hr. The data are presented as the percentage of the initial amount of ^{14}C-label added which was trapped as $^{14}CO_2$. Each data point represents the mean radioactivity in the nonsterile minus that observed in the sterile samples.

The effect of inorganic and organic nutrient amendments, pH, and temperature on benzene and toluene mineralization in contaminated and biostimulated material was determined by comparing the extent of mineralization in treated samples to that in reference samples incubated under in situ conditions (no amendments, 16°C, pH 8.0). All treatments were incubated at in situ temperature and pH unless stated otherwise. Treatments included incubation at or with: 26°C; pH 7.0; inorganic nutrients with 164, 16.4, or 1.6 mg/L N as NH_4Cl or 164 mg/L N as $NaNO_3$; 1, 10, and 100 mg/L humic acid (Aldrich Chemical, Milwaukee, WI); 1 mg/L acetate and inorganic nutrients with 164 mg/L N as NH_4Cl. The inorganic nutrients were added to achieve final concentrations per liter of ground water: 40 mg Na_2HPO_4, 60 mg KH_2PO_4, 20 mg $MgSO_4 \cdot 7H_2O$, 100 mg $NaCO_3$, 1 mg $CaCl_2 \cdot 2H_2O$, 2 mg $MnSO_4 \cdot H_2O$ and 0.5 mg $FeSO_4 \cdot 7H_2O$.

In a separate experiment, the effect of hydrogen peroxide on toluene mineralization in contaminated and biostimulated materials was determined. Microcosms containing 5 g of subsurface material were filled with sterile ground water amended with 10 mg H_2O_2/L. The extent of toluene mineralization in amended microcosms was compared to that in microcosms without peroxide. The concentration of dissolved oxygen in amended and unamended microcosms during the course of the experiment was determined also.

Statistical Analysis

The Statistical Analysis System (SAS Institute, Inc., 1982) was used to analyze the differences in microbial growth in different core materials and in biodegradation potential in materials amended with nutrients and incubated at different temperatures. The data were analyzed by analysis of variance using the ANOVA subprogram. Means were compared using the Scheffe test; all differences indicated as significant are at the 5% level. The difference between percent mineralization in H_2O_2 amended and unamended samples was compared using the Student's t test modified to reflect the unequal variances and reported at the 5% level unless indicated otherwise.

RESULTS

Site Characterization

The analysis of the three core materials indicates the presence of more carbon in the contaminated and biostimulated samples than in the uncontaminated samples (Table 1). Of interest is the high concentration of nitrate in the biostimulated sample. Although added during biostimulation (Minugh, 1987), phosphate was not detected in the biostimulated sample. The *in situ* temperature of the ground water was 16°C and the pH of the uncontaminated, contaminated, and biostimulated materials was 8.2, 8.0, and 8.1, respectively. The moisture content of these materials was 12.2, 10.9, and 8.0%, respectively.

TABLE 1 Analysis of Subsurface Material

Substance	Subsurface Material		
	Uncontaminated	Contaminated	Biostimulated
Total Carbon (%)	0.10(0.05)	0.93(0.09)	0.91(0.22)
Carbonate as C (%)	<0.016	0.73(0.12)	0.62(0.14)
Organic Carbon (%)	0.09(0.05)	0.22(0.06)	0.30(0.09)
Total Iron (%)	0.64(0.09)	0.48(0.04)	0.90(0.02)
Water Soluble Nitrate (mg/L)	<1	<1	45(6)
Water Soluble Nitrite as N (mg/L)	<1	<1	<1
Water Soluble Phosphate as P (mg/L)	<3	7(2)	<3

Quantification and Characterization of the Subsurface Microflora

Microbial numbers were significantly higher on Nutrient Agar spread with sample from the biostimulated zone than those on other types of media and from other zones (Table 2). Microbial counts from all three zones were not different when plated on the other types of solid media. An experiment designed to determine the potential for glucose mineralization at 1 ppm indicated that the compound was mineralized in samples from all zones and that the extent of mineralization (about 36%) after 7 days of incubation was not different between zones (data not shown).

TABLE 2 Viable Counts of Subsurface Microorganisms

Sample	Depth (ft)	Media		
		Nutrient Agar	Contaminated Ground Water Agar	Uncontaminated Ground Water Agar
		- - - - -Cells/g Dry Soil (SD) x 10^6 - - - -		
Uncontaminated	24-25.5	1.8(0.5)	1.6(0.1)	1.6(0.3)
Contaminated	24-25.5	1.5(0.6)	0.3(0.03)	0.8(0.4)
Biostimulated	24-25.5	65.5(9.2)	3.8(0.9)	0.5(1.5)

Biotransformation of BTEX

Biotransformation of toluene, ethylbenzene, m-xylene, and o-xylene was determined; analytical problems were encountered with benzene, therefore, quantitative biodegradation data are not presented. The results of the experiments are shown in Figure 2. Toluene, ethylbenzene, and m-xylene were rapidly biodegraded in all three core types, although the rate was generally faster in the biostimulated material. After 3 wk of incubation, these compounds were completely biodegraded in the biostimulated material, whereas some ethylbenzene and m-xylene persisted in the contaminated material and toluene persisted in the uncontaminated material. o-Xylene was recalcitrant in all subsurface materials. The biodegradation potential of the uncontaminated and contaminated materials was not significantly different.

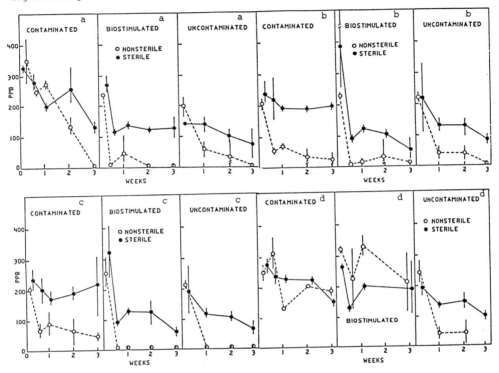

Fig. 2. Biotransformation of a) toluene, b) ethylbenzene, c) m-xylene, and d) o-xylene.

Mineralization

The mineralization potential of benzene and toluene was determined in the core materials after 1, 2, 3, 4, and 6 wk of incubation. Benzene mineralization was not detected in the uncontaminated material amended with 1, 10, 100, and 1000 µg benzene/L after 6 wk of incubation. Ambient levels of benzene measured by gas chromatography in the other core materials were below the level of detection (60 µg/L); 2 ng ^{14}C-benzene/L of ground water, was added to the microcosm. After 6 wk, 3.2% was mineralized in the contaminated samples and significantly more, 12.4%, was mineralized in the biostimulated samples. In contrast, toluene was mineralized in samples from all zones (Tables 3 and 4). After 6 wk of incubation, between 12 and 20% of the toluene was mineralized in the uncontaminated materials amended with 1 to 1000 µg/L and in contaminated and biostimulated materials which contained ambient toluene at 267 and 35 µg/L, respectively.

TABLE 3 Benzene Mineralization in Aquifer Material from Granger

Sample	Concentration (μg/L)	\multicolumn{6}{c}{Incubation Time (Wk)}

Sample	Concentration (μg/L)	1	2	3	4	5	6
		\multicolumn{6}{c}{-% Mineralization (SD)-}					
Uncontaminated	1,10,100,1000	0	0	0	0	0	ND[a]
Contaminated	0.002	ND	ND	ND	1.6(1.2)	ND	3.2(1.9)
Biostimulated	0.002	ND	ND	ND	8.4(2.2)	ND	12.4(6.2)

[a]Not determined

TABLE 4 Toluene Mineralization in Aquifer Material from Granger

Sample	Concentration (μg/L)	1	2	3	4	6
		\multicolumn{5}{c}{-% Mineralization (SD)-}				
Uncontaminated	1	0	1.5(0.1)	ND[a]	3.6[b]	16.5(4.5)
	10	2.0(0.9)	6.1(4.3)	10.6(1.6)	11.1	14.1(3.0)
	100	0	5.1(0.6)	4.9(0.8)	10.6(1.6)	ND
	1000	1.5(0.9)	6.3(0.5)	3.2(0.3)	6.9(3.2)	12.8(2.8)
Contaminated	267	ND	ND	18.2(2.5)	ND	14.3(3.4)
Biostimulated	35	ND	ND	22.2(6.4)	ND	20.2(12.3)

[a]Not determined
[b]Two replicates

Attempts were made to enhance the extent of benzene and toluene mineralization in samples from the contaminated and biostimulated zones which were incubated for 4 wk (Tables 5 and 6). For both zones, the extent of benzene mineralization in amended samples was not significantly different from that in unamended samples incubated at the *in situ* temperature and pH (reference samples). The extent of toluene mineralization in both zones was not significantly different from the reference sample, except for the acetate amendment in the contaminated material.

TABLE 5 Enhancement of Benzene and Toluene Mineralization in Contaminated Aquifer Material after 4 Wk of Incubation

	Compounds	
	Benzene	Toluene
	\multicolumn{2}{c}{-% Mineralization (SD)-}	
16°C, pH 8.0 (*in situ*)(Reference)	3.6(0.8)	27.9(6.1)
16°C, pH 7.0	1.6(0.5)	27.5(2.0)
26°C, pH 8.0	3.4(0.9)	20.7(7.5)
Humic acid		
1 mg/L	3.6(1.5)	36.5(9.5)
10 mg/L	3.4(0.6)	31.5(8.2)
100 mg/L	2.2(1.6)	23.5(2.9)
Inorganic nutrients with:		
164 mg/L N as NH_4Cl	1.8(0.5)	26.5(3.2)
16.4 mg/L N as NH_4Cl	2.4(0.5)	32.9(2.6)
1.6 mg/L N as NH_4Cl	2.2(0.5)	37.7(1.5)
164 mg/L N as $NaNO_3$	4.2(0.5)	28.7(2.7)
Inorganic nutrients with 1 mg/L acetate and 164 mg/L N as NH_4Cl	2.0(0.7)	45.7(8.2)[a]

[a]Significantly different from all treatments

TABLE 6 Enhancement of Benzene and Toluene Mineralization in Biostimulated Aquifer Material after 4 Wk of Incubation

Factors	Benzene	Toluene
	------ % Mineralization (SD) ------	
16°C, pH 8.0 (*in situ*)(Reference)	4.4(0.5)	8.6(3.9)
Humic acid		
1 mg/L	5.7(1.2)	8.8(1.5)
10 mg/L	5.2(1.0)	9.4(3.4)
Inorganic nutrients with		
164 mg/L N as NH_4Cl	4.7(1.7)	15.6(9.6)
16.4 mg/L N as NH_4Cl	3.3(0.8)	15.6(5.9)
Inorganic nutrients with		
1 mg/L acetate and 164 mg/L N as NH_4Cl	5.4(1.5)	5.1(2.0)

The effects of H_2O_2 amendment on toluene mineralization and dissolved oxygen concentration in contaminated and biostimulated samples are shown in Figure 3. The data indicate that the initial rate of toluene mineralization in the biostimulated material was significantly greater in amended samples (17.5%) than in unamended samples (8.1%). After 14 days of incubation, the H_2O_2 amended samples had slightly significantly ($\alpha=0.1$) greater mineralization (18.8%) than the unamended samples (9.0%). Dissolved oxygen was rapidly removed from microcosms containing amended and unamended biostimulated material, concomitant with toluene mineralization. Toluene was not mineralized in amended and unamended samples from the contaminated zone after 2 wk of incubation although some dissolved oxygen was consumed.

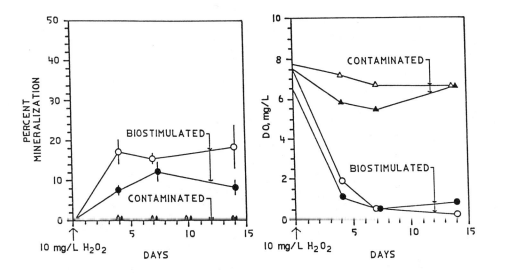

Fig. 3. Effects of H_2O_2 on toluene mineralization and dissolved oxygen (Open, H_2O_2 addition; Closed, no addition).

DISCUSSION

The results indicated that the subsurface microflora at the Granger site was active and could degrade many of the compounds found in gasoline. Microbial numbers and contaminant biodegradation potential were greater in samples from the biostimulated zone than in samples from the uncontaminated and contaminated zones. Microbial counts from the biostimulated zone on Nutrient Agar, 6.6×10^7 cells/g dry material, were not only higher than those in the uncontaminated and contaminated zones but also higher than most viable counts of subsurface microorganisms reported in the literature (Ghiorse and Wilson, 1988). The high cell counts detected in the biostimulated samples may have resulted from the addition of nutrients during the biostimulation demonstration. The contaminated and biostimulated materials contained about the same amount of carbon; nitrate was detected in the biostimulated sample only. Although ammonia was added during the biostimulation demonstration (Minugh et al., 1987), the presence of nitrate in the biostimulated zone may have resulted from nitrification of the ammonia amendment.

Benzene was mineralized in samples from the contaminated and biostimulated zones only, whereas toluene was mineralized in samples from all zones. In contrast to the small amount of benzene and toluene that was mineralized in these samples, a large amount of toluene, ethylbenzene, and m-xylene was biodegraded. Although the solvents used to add the ^{14}C-labeled compounds in the mineralization experiments and the unlabeled compounds in the biotransformation experiments were different, the discrepancy between the mineralization and biodegradation potentials may be a result of the initial incorporation of the compounds into cellular material rather than their metabolism through respiration. Swindoll et al. (1988a) have reported that several xenobiotics added to pristine subsurface samples were largely incorporated into microbial cells while only a small percentage of the compounds was mineralized.

Biodegradation of m-xylene but not o-xylene is contrary to some research indicating that aromatics with substituents in the ortho- and para- positions are more readily biodegraded than those with substituents in the meta position (Alexander and Aleem, 1961). However, Kuhn et al. (1985) and Kappeler and Wurhmann (1978) have reported that m-xylene is more biologically labile than o-xylene under aerobic conditions.

Attempts to enhance the extent of benzene and toluene mineralization with changes in environmental conditions and nutrient status were not successful, except for the addition of acetate which enhanced toluene mineralization in the contaminated material. Thomas et al. (1989) reported that organic and inorganic nutrient amendments did not significantly increase the mineralization potential of several polycyclic aromatic compounds in samples of contaminated ground water. In contrast, Swindoll et al. (1988b) found that the addition of inorganic nutrients enhanced the biodegradation of several xenobiotics in some, but not all, samples from a pristine aquifer, while the addition of glucose or amino acids inhibited biodegradation. In our experiments, the enhancement of toluene mineralization in the acetate-amended samples may have resulted from an increase in the number of toluene degrading organisms, although the amendment did not produce similar results for benzene. Other researchers have reported that an increase in biodegradation (acclimation) occurs as a result of an increase in the population of contaminant degraders (Wiggins et al., 1987). Also of interest is the greater mineralization potential of toluene in contrast to benzene. Higher removal efficiencies for toluene in comparison to benzene have been reported previously for methanogenic aquifer material amended with gasoline hydrocarbons (Wilson and Rees, 1985).

In peroxide amended samples of biostimulated material, the initial rate of toluene mineralization was significantly greater and the extent was slightly greater than in the unamended samples. No toxicity was associated with the H_2O_2 amendment; additionally, a stimulatory effect was observed in the biostimulated material. These results suggest that the toluene degrading organisms were tolerant of and adapted to the use of H_2O_2, the source of oxygen used in the field demonstration of in situ biorestoration. Toluene was not mineralized in the contaminated zone although some dissolved oxygen was consumed. The consumption of oxygen in the contaminated material may have been a result of toluene incorporation into cell material, as was suggested in our biotransformation experiments using gas chromatography which indicated rapid biological removal of toluene. These data also suggest that the mineralization potential was greater in the biostimulated material than in the contaminated material, possibly as a result of the higher cell density in the former.

The results of these experiments indicate that microbial numbers and activity in the biostimulated zone, which still contained residual hydrocarbon, remained enhanced for at least 2 yr after the in situ biorestoration process had been terminated. Similar results were reported for two successive field experiments conducted in the same test plot at Moffett Naval Air Station, Mountain View, California, which were designed to enrich for methanotrophs to cometabolize chlorinated solvents in ground water. Although about one year lapsed between field experiments, the subsurface microflora retained its ability to respond to methane addition and concomitantly cometabolize selected chlorinated solvents (Semprini, 1988).

DISCLAIMER

Although the research described in this article has been supported by the U. S. Environmental Protection Agency through Assistance Agreement No. CR-812808 to Rice University, it has not been subjected to Agency review and therefore does not necessarily reflect the views of the Agency and no official endorsement should be inferred.

REFERENCES

Alexander, M. and Aleem, M. I. H. (1961). Effect of chemical structure on microbial decomposition of aromatic herbicides. *Agric. Food Chem., 9*, 44-47.

Dunlap, W.J., McNabb, J.F., Scalf, M.R., and Cosby, R.L. (1977). Sampling for Organic Chemicals and Microorganisms in the Subsurface. EPA-600/2-77-176, R.S. Kerr Environmental Research Laboratory, Ada, OK.

Ghiorse, W.C. and Balkwill, D.L. (1983). Enumeration and characterization of bacteria indigenous to subsurface environments. *Dev. Ind. Microbiol., 4*, 213-224.

Ghiorse, W.C. and Wilson, J.T. (1988). Microbial ecology of the terrestrial subsurface. *Adv. Appl. Microbiol., 33*, 107-172.

Kappeler, T. and Wuhrmann, K. (1978). Microbial degradation of the water soluble fraction of gas oil--II. Bioassays with pure strains. *Water Res., 12*, 335-342.

Kuhn, E. P., Colberg, P. J. Schnoor, J. L. Warner, O., Zehnder, A. J. B., and Schwarzenback, R. P. (1985). Microbial transformations of distributed benzenes during infiltration of river water to ground water: Laboratory column studies. *Environ. Sci. Technol., 19*, 961-968.

Marinucci, A.C. and Bartha, R. (1979). Apparatus for monitoring the mineralization of volatile ^{14}C-labeled compounds. *Appl. Environ. Microbiol., 38*, 1020-1022.

Minugh, E., Ferguson, R., Smith, R., Raymond, R. L., Norris, R., and Brown, R. (1987). Field Study of Enhanced Subsurface Biodegradation of Hydrocarbons Using Hydrogen Peroxide as an Oxygen Source. API Publication No. 4448, American Petroleum Institute Washington, DC, 76 p.

SAS Institute, Inc. (1982). *SAS User's Guide: Statistics*, Cary, NC.

Semprini, L., Roberts, P.V., Hopkins, G.D., and McCarty, P.L. (1988). Field evaluation of aquifer restoration by enhanced biotransformation. In: *Proceedings of the International Conference on Physiochemical and Biological Detoxification of Hazardous Wastes;* 1988, Technomic: Lancaster, PA (in press).

Swindoll, C.M., Aelion, C.M., Dobbins, D.C., Jiang, O., Long, S.C. and Pfaender, F.K. (1988a). Aerobic biodegradation of natural and xenobiotic organic compounds by subsurface microbial communities. *Environ. Toxicol. Chem., 7*, 291-299.

Swindoll, C.M., Aelion, C.M., and Pfaender, F.K. (1988b). Influence of inorganic and organic nutrients on aerobic biodegradation and on the adaptation response of subsurface microbial communities, *Appl. Environ. Microbiol., 54*, 212-217.

Thomas, J.M., Lee, M.D., Scott, M.J., and Ward, C.H. (1989). Microbial ecology of the subsurface at an abandoned creosote waste site. *J. Ind. Microbiol., 4*, 109-120.

Thomas, J.M., Lee, M.D., Bedient, P.B., Borden, R.C., Canter, L.W., and Ward, C.H. (1987a). Leaking Underground Storage Tanks: Remediation with Emphasis on *In Situ* Biorestoration. EPA/600/2-87/008, U.S. Environmental Protection Agency, R.S. Kerr Laboratory, Ada, OK, .143 p.

Thomas, J.M., Lee, M.D. and Ward, C.H. (1987b). Use of ground water in assessment of biodegradation potential in the subsurface. *Environ. Toxicol. Chem., 6*, 607-614.

Thomas, J.M. and Ward, C.H. (1989). *In situ* biorestoration of organic contaminants in the subsurface. *Environ. Sci. Technol., 23*, 760-766.

Wiggins, B., Jones, S.H., and Alexander, M. (1987). Explanations for the acclimation period preceding the mineralization of organic chemicals in aquatic environments. *Appl. Environ. Microbiol., 53*, 791-796.

Wilson, B.H. and Rees, R.F. (1985). Biotransformation of gasoline hydrocarbons in methanogenic aquifer material. In: *Proceedings of the NWWA/API Conference on Petroleum Hydrocarbons and Organic Chemicals in Ground Water - Prevention, Detection and Restoration*, Houston, TX, National Water Well Association, Worthington, OH, pp. 128-139.

Wilson, J.T., McNabb, J.F., Balkwill, D.L. and Ghiorse, W.C. (1983). Enumeration and characterization of bacteria indigenous to a shallow water table aquifer. *Ground Water, 21*, 134-142.

ASSESSMENT OF THE POTENTIAL FOR *IN SITU* BIOTREATMENT OF HYDROCARBON-CONTAMINATED SOILS

Philip Morgan and Robert J. Watkinson

Shell Research Ltd., Sittingbourne Research Centre, Sittingbourne, Kent ME9 8AG, UK

ABSTRACT

Enhanced in situ biotreatment is a recent technology for the cleanup of contaminated soil and ground water but it has not yet been tested for many contaminants. This report describes the assessment of three hydrocarbon-contaminated sites, one contaminated with crude oil, one with lubricating oil and one with gasoline, with respect to their potential for biotreatment. All locations were permeable, sandy soils which contained low concentrations of extractable inorganic macronutrients. Degradative microbial populations were present, although their numbers were reduced in the most highly contaminated portions of the soil. Hydrocarbon analysis demonstrated that vertical penetration of contaminants into the soil was poor for the crude oil but had occurred at the other sites. There was some evidence that biodegradation at the crude and lubricating oil-contaminated sites may have occurred. The available data suggested that biotreatment of the lubricating and gasoline-contaminated sites by the provision of inorganic nutrients and oxygen to the soils might prove viable. However, it was found that the addition of inorganic nutrients resulted in an inhibition of mineralisation in the soils.

KEYWORDS

Biorestoration; biodegradation; cleanup; soil; land; oil; gasoline

INTRODUCTION

The use of enhanced in situ biodegradation to clean soils and ground water contaminated with organic compounds is a promising technology. However, its application to hydrocarbon contaminated sites has largely been limited to locations contaminated with gasoline or jet-fuels (Thomas et al, 1987; Lee et al., 1988, Morgan & Watkinson, 1989) and there is little information available concerning cleanup of heavier hydrocarbon contaminants. This report describes the analysis of three locations contaminated with different petroleum mixtures and conclusions are drawn as to the biotreatability of each.

SITES AND SAMPLING

Three sites were chosen to represent a range of hydrocarbon contaminants. All locations consisted of sandy soils with a relatively shallow ground water table. Site A was a former tanker (un)loading facility, site B was a former lubricating oil blending plant and site C was a gasoline station contaminated as a result of leakage from an underground storage tank. Soil cores were obtained by Begemann boring, sectioned in the laboratory and analysed for a variety of physicochemical and microbiological properties.

SOIL PHYSICOCHEMICAL PROPERTIES

All sites consisted of medium to fine sand and should therefore be of suitable permeability for a treatment programme. All are of a pH suitable for biological activity and the soils generally contain low concentrations of extractable inorganic nutrients and non-petrogenic organic carbon (Table 1).

TABLE 1 - Basic Physicochemical Properties of Soil Samples

Parameter	Site A	Site B	Site C
pH	7.5-8.3	8.0-8.7	6.5-7.5
Water table (m)	1.2-1.5	2.0	1.5-2.0
Total organic carbon (%)	0.5-2.0*	1.0-3.0	0.2-2.0
NO_3^- and NH_4^+ (μg ml^{-1})	2-10	2-8	2-20
Phosphate (μg ml-1)	0.1-4	0-2	0-10

* = Significantly higher (6-8%) within the top 25 cm of soil

ANALYSIS OF CONTAMINATION

The hydrocarbon content of soils A and B was ascertained by extraction and gravimetric determination. Extracted hydrocarbon was also fractionated into aliphatic, aromatic and polar components by column chromatography on silica gel 60 (Merck) - Woelm acid alumina (ICN Biomedical). The aliphatic fraction was eluted with hexane, the aromatic fraction with benzene and the polar fraction with 1:1 benzene-methanol. Aliphatics were further analysed by capillary gas chromatography (GC) on CP-SIL-5-CB (50-300°C at 5°C min^{-1}) and aromatics by high performance liquid chromatography (HPLC) on an ODS2 column using an eluent gradient of 60:40 acetonitrile-water to 100% acetonitrile.

The hydrocarbon content in soil A is illustrated in Figure 1a. The extracted material was black and tar-like and poor vertical penetration was noted. The hydrocarbon mixture was apparently enriched in polar components which may be indicative of biodegradation having occurred (Fig. 1b). Examination of GC traces of the aliphatic fraction demonstrated the predominance of heavy compounds and an unresolved mixture was present. Analysis of the aromatics demonstrated that high molecular weight compounds predominated. These data may suggest that degradation has resulted in the proportional accumulation of higher molecular weight components and potentially increased recalcitrance.

The hydrocarbon content of soil B is illustrated in Figure 2a. The extracted material was a light mobile oil and it was found to have penetrated to the ground water table where it was present as a free phase. The relative proportions of components is illustrated in Figure 2b and it can be seen that the hydrocarbon is primarily aliphatic. GC analysis of the oil revealed that in the upper regions of the soil there was a predominance of high molecular weight and unresolved components suggesting that biodegradation may have occurred.

Soil C was analysed for benzene, toluene and xylenes (BTX) by headspace gas chromatography. High concentrations of BTX were present close to the tank and one example set of data is illustrated in Figure 3. GC analysis showed no significant difference between the hydrocarbon composition of the soil and that of gasoline samples suggesting that little biodegradative or evaporative loss had occurred.

MICROBIAL POPULATIONS

All sites had relatively low microbial populations as determined by acridine orange-stained direct counts (AODC) and by cultural techniques employing mineral salts media supplemented with crude oil (sites A and B) or incubated in the presence of gasoline vapour (site C). Typical AODC counts per gram of soil were 10^8 in soil A and 10^7 in soil C. Significant numbers of aerobic hydrocarbon-degrading organisms were cultivated from all soils with the exception of the most highly contaminated regions of soil C where a marked depletion in microbial population was evident, presumably owing to solvent-type toxicity of the contaminant to microbial cells. Nevertheless, even in these cases approximately 10^3 gasoline-degrading cells were present per gram of soil.

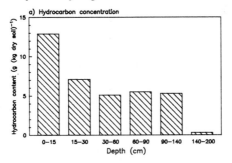

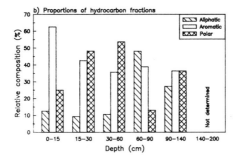

Fig.1. Hydrocarbon content through representative core of soil A. a) Total hydrocarbons by gravimetric determination. b) Relative proportions of hydrocarbon fractions.

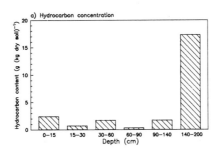

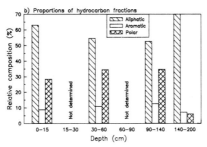

Fig.2. Hydrocarbon content through representative core of soil B. a) Total hydrocarbons by gravimetric determination. b) Relative proportions of hydrocarbon fractions.

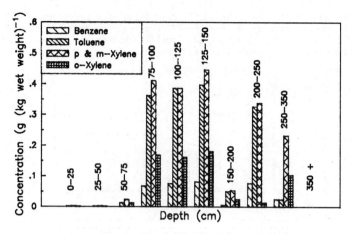

Fig. 3. Concentrations of benzene, toluene and xylene in representative core of soil C.

The potential for biodegradation of the contaminants in all of the soils was therefore demonstrated by the presence of adapted populations and the evidence for degradation having occurred in soils A and B. It was felt that site A would provide the slowest degradation rate owing to the recalcitrant nature of the hydrocarbon present and further experiments were conducted to attempt to optimise degradation rate by the provision of inorganic nutrients in simple laboratory batch cultures. In view of the requirement for a rapid screen of the potential activity of the soil, samples (5g) were placed into glass vials, supplemented with ^{14}C-radiolabelled phenanthrene, anthracene, naphthalene or hexadecane, sealed with teflon-lined stoppers and incubated at 20°C. At intervals, samples were sacrificed for analysis. 50% H_3PO_4 was injected into the vials to produce a slurry and the samples were sparged with CO_2-free air for 20 minutes. CO_2 in the off-gas was trapped in Carbosorb (Canberra Packard) and radioactivity determined by liquid scintillation counting. It was found that fertilisation did not increase degradation rate and some supplements resulted in a significant inhibition of mineralisation. Example data for naphthalene mineralisation in soil A are illustrated in Table 2. No alteration in pH was induced by these fertiliser additions.

TABLE 2 - Effect of Inorganic Nutrient Addition Upon Naphthalene Mineralisation in Soil A

Nutrient addition (per g soil)	Mineralisation rate (% of untreated)
10 mg KNO_3	25.5*
5 mg NH_4NO_3	20.1*
4 mg NaH_2PO_4	71.2
10 mg NH_4NO_3 + 1.5 mg NaH_2PO_4	21.0*
5 mg NH_4NO_3 + 0.8 mg NaH_2PO_4	79.2
1 mg NH_4NO_3 + 0.2 mg NaH_2PO_4	99.3
2 mg urea + 1 mg NaH_2PO_4	48.2*

* = significantly different from untreated control at p<0.01.

This inhibitory effect was noted not only for hydrocarbon substrates but was also apparent when glucose was the carbon source provided (Figure 4). The reasons for these observations are unclear but it may be that the addition of nutrients resulted in increased incorporation of carbon into biomass rather than mineralisation. Alternatively, the organisms present in the soil have become strongly adapted to an oligotrophic (low nutrient) life style, particularly with respect to inorganic nutrient supply. It is known that oligotrophic organisms can be inhibited by nutrient supplementation (Poindexter, 1981; Morgan & Dow, 1986) and this may be produced by exposure to organic or inorganic nutrients. It is evident that caution may be necessary in using nutrient supplements at contaminated sites if enhanced activity of indigenous organisms is intended.

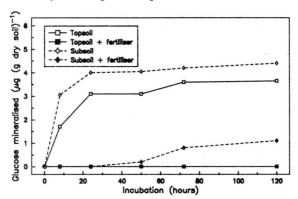

Fig. 4. Glucose mineralisation in soil samples from site A as a function of fertiliser (10 mg NH_4NO_3 + 1.5 mg NaH_2PO_4 per gram of soil) addition. Topsoil is from 0-25 cm depth and subsoil from 150-200 cm depth.

POTENTIAL FOR SITE BIOTREATMENT

The contamination at site C was potentially readily degradable but a preliminary venting programme would be a cost-effective means of removing the bulk of the contamination present. Biotreatment could then be used as a final "polishing" stage.

Site B would apparently be suitable for biotreatment following the pumping of free oil from the ground water table. Evidence for biodegradation at this site was obtained and the contamination appeared degradable by in situ organisms.

At site A it appeared that the residual material consisted primarily of branched aliphatic, high molecular weight aromatic and polar compounds. It may have been that significant biodegradation had resulted in proportional enrichment of these materials. It was noted that the indigenous microorganisms appeared to be oligotrophic and were therefore inhibited by high concentrations of inorganic nutrient supplements. This location would appear to be more suitable for a physicochemical treatment process.

CONCLUSIONS

This report demonstrates that hydrocarbon-contaminated sites may differ greatly in terms of both microbiological and physicochemical parameters. Previous biodegradation at long-standing sites may result in increased proportions of components of greater recalcitrance. Furthermore, preferential biodegradation of "simpler" components may give a rapid initial degradation rate but this will slow with time. The question of target concentrations (total hydrocarbon or individual components) and sustainable degradation rates must be addressed in order to obtain a realistic time-scale for treatment. It is also important to note that soil microorganisms may behave as oligotrophs and that assessment of fertiliser effects is necessary prior to application. It is therefore vital to assess a site for proposed biological treatment in great detail prior to cleanup and to test the impact of the chosen methodology upon the microbial population

REFERENCES

Lee, M.D., Ward, J.M., Borden, R.C., Bedient, P.B., Ward, C.H. and Wilson, J.T. (1988). Biorestoration of aquifers contaminated with organic compounds. CRC Crit Rev. Environ. Control 18, 29-89.

Morgan, P. and Dow, C.S. (1986). Bacterial adaptations for growth in low nutrient environments. In: <u>Microbes in Extreme Environments</u>, R.A. Herbert and G.A. Codd (Eds). Academic Press, London, pp 187-214.

Morgan, P. and Watkinson, R.J. (1989). Hydrocarbon degradation in soils and methods for soil biotreatment. <u>CRC Crit. Rev. Biotechnol.</u> <u>8</u>, 305-333.

Poindexter, J.S. (1981). Oligotrophy. Fast and famine existence. <u>Adv. Microb. Ecol.</u> <u>5</u>, 63-89.

Thomas, J.M., Lee, M.D., Bedient, P.B., Borden, R.C., Canter, L.W. and Ward, C.H. (1987). <u>Leaking underground storage tanks: remediation with emphasis on in situ biorestoration</u>. US EPA report no. EPA/600/2-87/008. Robert S. Kerr Environmental Research Laboratory, Ada, Oklahoma.

ENZYMATIC OXIDATION OF SOME SUBSTITUTED PHENOLS AND AROMATIC AMINES, AND THE BEHAVIOUR OF SOME PHENOLOXIDASES IN THE PRESENCE OF SOIL RELATED ADSORBENTS

H. Claus and Z. Filip

Institut für Wasser-, Boden- und Lufthygiene des Bundesgesundheitsamtes, Aussenstelle Langen, Paul-Ehrlich-Strasse 29, D–6070 Langen, FRG

ABSTRACT

When considering phenoloxidases as agents in treating chemical pollution in soil and groundwater environments, possible effects of the specific soil constituents and environmental factors on these enzymes should be investigated. In our in vitro experiments various aromatic amines and alkyl or halogensubstituted phenols were oxidized by a laccase and a tyrosinase. Oxidation of the latter compounds was accompanied by partial dehalogenation. However, the activity of phenoloxidases was differently inhibited by the presence of clays and clay-humus complexes. Also pH and temperature exerted different effects on the phenoloxidases.

KEYWORDS

Xenobiotics; phenoloxidases; adsorbents.

INTRODUCTION

Different xenobiotics may persist in the soils and underground environments or they may be partly or completely degraded by microorganisms or abiotic mechanisms. Other reactions, which aromatic compounds may undergo in nature are enzymatic coupling, catalyzed by extracellular oxidoreductases mainly of microbial origin. These include phenoloxidases such as laccases, tyrosinases, and peroxidases. The oxidation products are oligomers or polymers which become covalently linked to the soil organic matter.

Experimental results demonstrate a possible use of phenoloxidases in removing phenols from waste water (Atlow et al., 1984; Duguet et al., 1986), drinking water (Maloney et al., 1986) and for decontamination of polluted soils (Berry and Boyd, 1985; Dec and Bollag, 1988). In order to elucidate the possibility of enzyme application to polluted environments in more detail, we investigated the behaviour of phenoloxidases in the presence of different substrates, some abiotic soil factors and solid constituents (pH, temperature, quartz, clays).

MATERIALS AND METHODS

Laccases were prepared from the fungi *Polyporus versicolor* and *Pleurotus ostreatus* and a tyrosinase from *Streptomyces eurythermus* in the laboratory.

Table 1 Some Characteristics of the Phenoloxidases Used in the Tests

Enzyme	Molecular Weight (K Dalton)	IEP	Number of Isoenzymes	Other Features
Laccase (Polyporus versicolor)	61	3,5	2	Inducible glycoprotein contains Cu
Laccase (Pleurotus ostreatus)	47	3,5	3 - 6	Inducible glycoprotein contains Cu
Peroxidase Horseradish	40	5,3 - 7,3	5	Hemoprotein
Tyrosinase (Agaricus bisporus)	110	4,8	Tetramer	Contains Cu
Tyrosinase (Streptomyces eurythermus)	16	7,4	1	Contains Cu

Table 2 Clays Used for Adsorption Studies

Clay	Characteristics	C E C [meq/100 g]
3-layer clays		
B 1: Bentonite	Native aluminium-silicate	81
B 2: Bentonite	Natural Ca-bentonite from Hallertau (Bayern) min. contents of montmorillonite 90 %	96
B 3: Bentonite	As above but min. contents of montmorillonite 60 - 65 %	61
2-layer clays		
K 1: Kaolin	Essentially kaolinite particle size > 40 µm ca. 15 % > 20 µm ca. 40 % < 20 µm ca. 60 %	6,4
K 2: Kaolinite	Well crystallized (KG al, Georgia) specific surface $10,05 \pm 0,02\ m^2/g$	4,7

Tyrosinase from _Agaricus bisporus_ and peroxidase from horseradish were commercial preparations (Table 1). Bentonites and kaolinites, soil clays of different origin and cation exchange capacities, were used in the experiments in order to elucidate the effects of these specific soil and underground constituents (Table 2). Clays were also saturated with individual cations or with a humic acid preparation. Quartz sand and porous glass served as control silicate materials with low cation exchange capacities and adsorption activities. Enzyme activities with different substrates were determined spectrophotometrically or by estimating oxygen consumption or chloride release electrometrically. Details of methods have been described elsewhere (Claus and Filip, 1988).

RESULTS AND DISCUSSION

Behaviour of selected substrates in the presence of phenoloxidases

The oxidation of 49 phenols and aromatic amines by phenoloxidases was tested (Table 3). Thirty-eight compounds (78 %) were oxidized by the laccase from _P. versicolor_ and 7 (14 %) by tyrosinase from _A. bisporus_. Anisol was the only substrate which was weakly oxidized by tyrosinase and not by laccase. Alkyl-substituted phenols were rapidly oxidized by laccase. No reaction occurred with bulky molecules such as 2,6-di-tert-butyl-4-methylphenol, probably because of some steric hindrance. Alkyl-chloro-substituted phenols (7 compounds) were oxidized by laccase. Of 16 halogenated phenols, including F and Br containing substitutes, 11 were oxidized by the laccase. Reaction velocity decreased with the increasing degree of halogen substitution. No oxygen consumption was observed with pentachlorophenol. No reaction occured with nitro-substituted phenols. No synergistic reactions were observed when laccase and tyrosinase were added simultaneously to substrate solutions. With para-substituted phenols, reaction velocity decreased in the sequence: amino > isopropyl > fluoro > chloro > nitro residues. Substituents in meta-positions, e.g., 3-chlorophenol and 3-aminophenol, mostly exerted negative effects on reaction velocity.

Oxidation of chlorinated compounds was often accompanied by a different degree of dehalogenation. Using tyrosinase, chloride became released from some substrates even in the absence of oxygen consumption (Table 4).

Stability of phenoloxidases at different pH and temperature

The stability of phenoloxidases showed a significant dependence from the pH. Laccases were active between pH 2.0 and 10, with an optimum between pH 5.0 and 6.0. Peroxidase was strongly inactivated at pH < 4.0 but its activity remained nearly constant above pH 4.0. The tyrosinase stability ranged from 4.0 to 10 for the preparation from _A. bisporus_ and from 5.0 to 10 for the preparation from _S. eurythermus_. The optimum activity was observed at pH 7.0 or slightly above this value.

When kept for 14 days at 4 °C and the optimum pH, laccases showed no activity losses, whereas other phenoloxidases under test retained only about 70 % of their original activity. All preparations of phenoloxidases and especially the tyrosinases were more stable in native or autoclaved groundwater at 10 °C than in distilled water (Table 5).

The resistance of tyrosinase from _A. bisporus_, either free or adsorbed on kaolinite at pH 5.0, was tested at temperatures elevated up to 65 °C (Table 6) and in the presence of proteolytic enzymes (Table 7). Adsorbed tyrosinase was found to be more sensitive.The specific surface acidity and/or thermal accumulation capacity of the kaolinite may account for the lower thermal resistance of the tyrosinase adsorbed on the surface of this clay mineral.

Adsorption of phenoloxidases

The adsorption of phenoloxidases on clays was strongly enhanced by decreasing pH. In the presence of bentonites, laccase activity was rapidly reduced at pH < 5.0 and disappeared completely at pH 3.0, indicating that all the enzyme was

Table 3 Oxidation of Some Substituted Phenols and Aromatic Amines by a Laccase and a Tyrosinase

Compound	Oxygen Consumption (nMol O_2 min^{-1} ml^{-1})	
Alkyl-Substituted Phenols	Laccase	Tyrosinase
2,6-dimethylphenol	9600	1800
2,4,6-trimethylphenol	5200	0
2,3,5-trimethylphenol	3000	0
2,3,6-trimethylphenol	3000	0
2-isopropylphenol	3700	1400
3-isopropylphenol	4100	0
4-isopropylphenol	4400	0
2,6-diisopropylphenol	5200	0
2,4-di-tert.butylphenol	700	0
2,5-di-tert.butylphenol	0	0
2,6-di-tert.butyl-4-methylphenol	0	0
Substituted Chlorophenols		
2-chloro-6-methylphenol	6600	0
2-chloro-5-methylphenol	2200	0
4-chloro-2-methylphenol	3500	0
4-chloro-3-methylphenol	1500	0
2-chloro-4,5-dimethylphenol	6600	0
2,4-dichloro-6-methylphenol	5900	0
2,2-methylen-bis-3,4,6-trichlorophenol	400	0
Halogen-Substituted Phenols		
2-chlorophenol	3800	0
3-chlorophenol	400	0
4-chlorophenol	900	0
2,4-dichlorophenol	1600	0
2,6-dichlorophenol	2500	0
2,3-dichlorophenol	200	0
3,4-dichlorophenol	0	0
2,5-dichlorophenol	0	0
2,4,6-trichlorophenol	2300	0
2,4,5-trichlorophenol	300	0
2,3,4-trichlorophenol	0	0
2,3,5-trichlorophenol	0	0
2,3,5,6-tetrachlorophenol	700	0
pentachlorophenol	0	0
4-fluorophenol	1600	0
4-bromo-2-chlorophenol	2200	0
Arylamines, Nitrophenols		
2-aminophenol	41000	15000
3-aminophenol	3100	0
4-aminophenol	8500	900
2,5-dimethylaniline	2600	0
N-methylaniline	4400	0
N,N-dimethylaniline	5900	0
3-aminobenzo-trifluorid	400	0
2-nitrophenol	0	0
3-nitrophenol	0	0
4-nitrophenol	0	0
Other Phenols		
phenol	1000	8800
2-phenylphenol	600	0
4-methoxyphenol	3700	28000
4-(2-phenyl-2-propyl)-phenol	1700	0
anisol	0	400

Table 4   Dehalogenation of Substituted Phenols by Some Phenoloxidases

PHENOL[1]	CL⁻ RELEASE BY PEROXIDASE MG/L	%	LACCASE MG/L	%	TYROSINASE MG/L	%
2-CHLOROP.	13	1	9	1	9	1
4-CHLOROP.	31	15	17	9	58	29
2,6-DICHLOROP.	18	7	46	19	0	0
2,4,5-TRICHLOROP.	44	-	30	-	0	-
2,4,6-TRICHLOROP.	100	-	77	-	0	-
2,3,5,6-TETRACHLOROP.	9	-	0	-	0	-
2-CHLORO-3-METHYLP.	26	-	45	-	17	-
4-CHLORO-2-METHYLP.	58	23	99	39	11	4
4-CHLORO-3-METHYLP.	33	23	19	14	66	47
2-CHLORO-4,5-DI-METHYLP.	25	-	24	-	0	-
2,4-DICHLORO-6-METHYLP.	54	-	40	-	0	-
4-BROMO-2-CHLOROP.	60	64	36	38	0	-

1) CONCENTRATION: 500 - 6000 MG/L
%: CL⁻ RELEASED/CL⁻ BOUND

adsorbed. In the presence of kaolinites the decrease of free laccase activity was slower and some activity still remained at pH 2.0. Similar results were obtained for the other enzymes. Maximum adsorption occurred at pH values near the isoelectric points (IEPs) of the phenoloxidases.

Enzyme adsorption correlated well with the cation exchange capacity (CEC) of clays: at the adequate pH, phenoloxidases were strongly adsorbed on bentonites and a bentonite-humus complex, often resulting in total inhibition of enzyme activity. The kaolinites were less effective. Quartz sand exerted similar effects as kaolinites, whereas porous glass had no effect (Figure 1).

The binding capacities of the homoionic clays were estimated from the adsorption isothermes: About 0.5 mg peroxidase and 1-2 mg tyrosinase (from A. bisporus) were adsorbed by 1.0 mg bentonite. Only 10 µg tyrosinase (from A. bisporus) were adsorbed by 1.0 mg kaolinite. All phenoloxidases but tyrosinase from A. bisporus lost their activity when adsorbed on bentonite. The bound tyrosinase retained between 2% and 11% of the free enzyme activity. With kaolinite, the retained activity was between 20% and 50% for tyrosinase from A. bisporus and about 15 % for laccase from P. versicolor.

Attempts were made to desorb the complexed phenoloxidases by increasing pH and ionic strength (Figure 2). The enzymes were found to be held more strongly by bentonite than by kaolinites: above the IEP of the individual enzyme about 45% of the tyrosinase activity became desorbed from the kaolinites but only 15% from the bentonites. An increase in the ionic strength from 0.2 to 1.0 M NaCl did not influence desorption of the enzymes. Specific activities of desorbed phenoloxidases were distinctly reduced as compared with the controls.

Table 5   Half-life Times (Days) of Phenoloxidases in Groundwater at 10 °C

ENZYME	GROUNDWATER[1] AUTOCLAVED	GROUNDWATER[1] NATIVE	DIST. WATER[2]
TYROSINASE FROM S. EURYTHERMUS	57	37	11
TYROSINASE FROM A. BISPORUS	>92	>92	23
LACCASE FROM P. OSTREATUS	15	10	6
LACCASE FROM P. VERSICOLOR	19	29	7
PEROXIDASE FROM HORSERADISH	21	>92	>92

1) pH 6.95   2) pH 5.50

Table 6   Stability of Free and Kaolinite-Adsorbed Tyrosinase at Different Temperatures

TIME	°C	% ACTIVITY FREE	% ACTIVITY ADSORBED
24 h	4	100	100
24 h	28	64	10
15'	50	52	34
15'	55	21	2
15'	60	1	0
15'	65	0	0

Table 7   Stability of Free and Kaolinite-Adsorbed Tyrosinase Against Proteolytic Enzymes

ENZYME	T (h)	% ACTIVITY FREE	% ACTIVITY ADSORBED
CONTROL	2	100	100
PRONASE E	2	32	0
PRONASE E	4	2.3	0
PROTEINASE S	2	100	43
PROTEINASE S	4	95	6.9

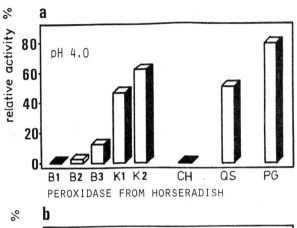
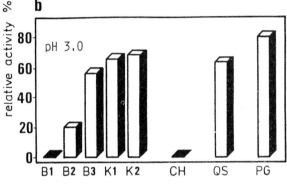
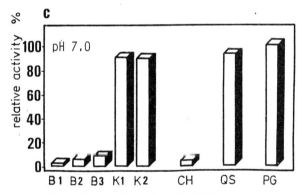

B: Bentonite   K: Kaolinite   CH: Clay-Humus Complex
QS: Quartz Sand
PG: Porous Glass

Fig. 1. Residual enzyme activities after adsorption.

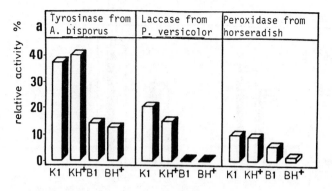

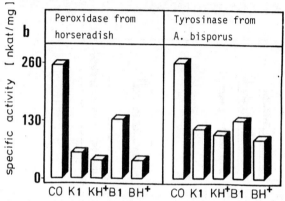

CO: Control   K: Kaolinite   B: Bentonite

Fig. 2.   Enzyme activities released at pH 8.0.

The results of these investigations may be summarized as follows:
- Phenoloxidases have been found to oxidize a great number of substituted phenols and aromatic amines
- Phenoloxidases are relatively stable in a broad pH and temperature range
- Phenoloxidases specifically adsorb on clays and clay-humus complexes
  - The adsorption correlates positively with the CEC of the clays
  - The highest adsorption occurs at pH values near to the IEP of the enzymes
  - Phenoloxidases adsorbed on clays and clay-humus complexes partly lose their activity

In conclusion, the broad substrate spectrum of phenoloxidases and their resistance against environmental stress caused by pH, and temperature changes indicate that some of these enzymes, and especially laccases and peroxidases, can be considered as possible agents to treat some chemical pollutants in different soil and underground environments.

ACKNOWLEDGEMENT

This research was supported by a post-doctoral stipend granted by courtesy of BMFT Bonn to Dr. H. Claus.

REFERENCES

Atlow, S.L., Bonadonna, A.L. and Klibanov, A.M. (1984). Dephenolization of industrial waste water catalyzed by polyphenol oxidase. Biotechnol. Bioeng., 26, 599-603.

Berry, D.F. and Boyd, S.A. (1985). Decontamination of soil through enhanced formation of bound residues. Environm. Sci. Techn., 19, 1132-1133.

Claus, H. and Filip, Z. (1988). Behaviour of phenoloxidases in the presence of clays and other soil related adsorbents. Appl. Microbiol. Biotechn., 28, 506-511.

Dec, J. and Bollag, J.M. (1988). Microbial release and degradation of catechol and chlorophenols bound to synthetic humic acid. Soil Sci. Am. J., 52, 1366-1371.

Duguet, J.B., Dussert, B., Bruchet, A. and Mallevialle, J. (1986). The potential use of ozone and peroxidase for removal of aromatic compounds from water by polymerization. Ozone Sci. Eng., 8, 247-260.

Maloney, S.W., Manem, J., Mallevialle, J. and Flessinger, F. (1986). Transformation of trace organic compounds in drinking water by enzymatic oxidative coupling. Environm. Sci. Techn., 20, 249-253.

BASIC STUDY ON TCEs BEHAVIOR IN SUBSURFACE ENVIRONMENT

K. Muraoka* and T. Hirata**

*Department of Civil Engineering, Osaka University, 2-1 Yamada-oka, Suita City, Osaka-565, Japan
**Division of Water and Soil Environment, The National Institute for Environmental Studies, 16-2 Onogawa, Tsukuba City, Ibaraki-305, Japan

ABSTRACT

In order to clarify the aspects of initial stage of groundwater pollution due to organochlorines in the subsurface, some experiments on the movement of contaminants near the groundwater surface and solution process into groundwater were conducted. As one result, the oscillation of water level is assumed to support the solubility and the durability of contamination of groundwater. As the second subject of this work, the monitoring in a model region was pursued for the purpose of explaining the distribution of contamination. The method of soil gas analysis was then used to investigate the state of groundwater pollution.

KEYWORDS

Groundwater pollution, Trichloroethylene, Tetrachloroethylene, Dissolution, Monitoring, Soil gas

GROUNDWATER POLLUTION BY TCES IN JAPAN

Since The Environment Agency of Japan opened in 1983 the data of groundwater pollution by TCEs (Trichloroethylene, Tetrachloroethylene and 1,1,1-Trichloroethane) in ten large cities and five local cities, people in many regions have considered this pollution very serious. Even recently new events have been often reported here and there. According to the official information of local governments, the condition of groundwater pollution from 1984 to 1986 is shown in Table 1. Organochlorines were detected approximately in a third part of the groundwater samples; moreover about 3% of the samples with trichloroethylene overshot the provisional standard for drinking water (0.03 mg/l).

The field surveys have suggested that serious groundwater pollution due to the organochlorines was often caused by the intrusion of the undiluted liquids into the subsurface environment. As the the undiluted organochlorines have such different properties from water, their basic behavior characteristics, such as the migration and leaching in the soil and groundwater environment, should be experimentally studied.

A huge amount of the organochlorines are still produced worldwide; they are consumed to clean up the metals, semi-conductor, hi-tech manufactures, clothes etc. which are deeply associated with daily life of people. Such broad utility of organochlorines makes it difficult to identify the pollutant

Table 1 Groundwater pollution by TCEs in Japan (1984-1986)

	Contaminant	Number of city,town	Number of observed wells	Number of wells over standards	Ratio of excess (%)
1984	TCE	833	5720	122	2.1
	PCE		5733	185	3.2
	TCA		5476	4	0.1
1985	TCE	468	3461	123	3.6
	PCE		3459	140	4.0
	TCA		3455	8	0.2
1986	TCE	303	2794	146	5.2
	PCE		2777	109	3.9
	TCA		2763	3	0.1
Total	TCE	—	11975	391	3.3
	PCE	—	11969	434	3.6
	TCA	—	11694	15	0.1

note 1)TCE:Trichloroethylene, PCE:Tetrachloroethylene, TCA;1,1,1-trichloroethane.
2)Observed wells are different in each year.
3)Provisional standards as to drinking water are TCE:0.03mg/l, PCE:0.01mg/l and TCA:0.3mg/l (less than).

source in the polluted region, because the latent pollutant sources are located here and there in the specified region and in addition the groundwater flow system is not clear. The first stage of countermeasures in these circumstances is boring and sampling the groundwater to uncover the details of groundwater pollution.Therefore a time-saving and less expensive monitoring technique with high accuracy is required to determine the boundary of the polluted area and the pollutant sources in the local groundwater network. This paper describes the basic experimental results concerning the migration mechanism of the pollutants and the monitoring technique using analysis of surface soil gas of volatile contaminants from subsurface environment.

BASIC EXPERIMENT FOR SOLVING THE MECHANISM OF GROUNDWATER POLLUTION

Migration of undiluted trichloroethylene in subsurface environment

Trichloroethylene (TCE for short hereafter) infiltrates readily into the unsaturated soil, permeates into the soil aggregation then remains there. TCE can infiltrate even in the saturated soil when the pore size in soil is larger than around 3mm (Hirata and Muraoka, 1988), but it is generally hard to keep going down in the saturated media and finally is stagnant and remains there.

When the pore space is small, below 1mm, TCE liquid cannot intrude into the saturated zone and remains stagnant on the groundwater surface.In the case that the groundwater level keeps stationary, TCE liquid never moves on it, however, it starts to diffuse in the porous media when the groundwater level goes up and down. Fig. 1 shows the migration of the TCE liquid in the porous media driven by the oscillation of the groundwater surface. Experiments were permormed in a cylindrical column 5cm in diameter and 50cm in depth with the unsaturated portion in the upper half and saturated portion in the lower half. The colored trichloroethylene as the test liquid was placed on the surface of the porous media by a syringe, and its

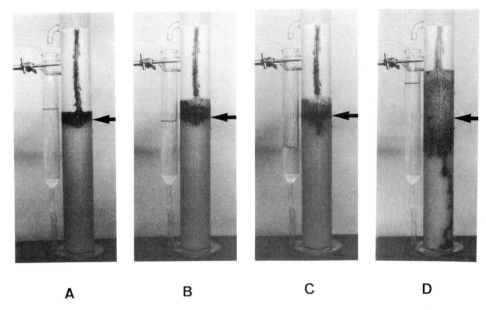

Fig. 1 Migration of trichloroethylene in porous media driven by oscillation of groundwater surface. Oscillation was produced by in- and out-put of water in a small glass tube connected with the cylindrical column. Up and down velocity of water surface is approximately 1.5 cm/min. Arrows show the position of initial groundwater surface.

flowing down the wall of the measuring cylinder was observed. The volume of test liquid was 10 ml.

As the mean diameter of the particles is small, the TCE liquid remains stagnant on the water surface after the intrusion in unsaturated zone (see Fig. 1 A). Then during the flood time, in which groundwater level is rising slowly, the TCE liquid seldom moves, while only water volume occupies the upper part over the TCE layer (see Fig. 1 B). During the ebb time when water level is falling, TCE liquid starts to intrude into the lower aqueous layer as TCE is dispersing into small lumps (Fig. 1 C). Fig. 1 D shows the final state of the experiment after two cycles of oscillations. The oscillation of the groundwater level promotes the diffusion of TCE downward in the saturated porous media, while there is no movement of TCE if water level keeps constant.

TCE leaching due to rainfall infiltration

TCE leaching from the soil due to the rainfall infiltration was examined by using the cylindrical column as shown in Fig. 2. The porous medium was constituted by the Kanuma soil of which aggregations have approximate 1.5 mm of mean diameter. Pure water falls constantly on the top of the soil column with 10 mm/hr of rainfall intensity. Fig. 3 shows the time variation of concentration of percolating water from the bottom of test cylinder and the concentration at early stage of time indicates 1000 mg/l which is nearly equal to the maximum solubility of TCE (1100 mg/l) in the water. Total amounts of leached TCE in both cases come up to about 70 % of injected TCE.

TCE solution by water flow both in two layers and in porous media

Fig. 2 Cylindrical column (60φ×200mm) for leaching test

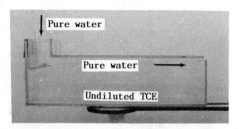

Fig. 4 Rectangular tube (200×50×50mm) for dissolution test

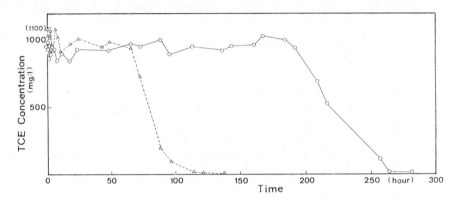

Fig. 3 TCE leaching from unsaturated soil due to rainfall infiltration. Symbols show the different amounts of TCE injected in the soil column; circle = 6g and triangle = 2g.

Two kinds of the elution experiment were conducted by using a rectangular glass tube shown in Fig. 4. One is the elution test in two layers of which upper layer makes pure water flow and lower layer consists of undiluted trichloroethylene liquid. The other is the elution test in the porous media in which flow in the upper part is by pure water and in the lower part by undiluted trichloroethylene liquid. Fig. 5 shows the relationship between the concentration and velocity of the upper layer in two cases. It is clearly recognized that the TCE concentration of the upper layer decreases with increasing of the upper layer velocity.

Fig. 6 illustrates the elution coefficient E against the particle Reynolds number Re_d, where $E = U_e/U$; U_e : mixing velocity of TCE between two layers; U : real velocity of the upper layer and $Re_d = Ud/\nu$; d : mean diameter of the glass beads; ν : kinematic viscosity. In this normalization, the effect of the pore space seems to vanish. Plots can be approximately presented by a linear line on log-log scale, therefore, the elution coefficient E is presented in the following form using the particle Reynolds number.

$$E = 7.41 \times 10^{-6} Re_d^{-0.515} \tag{1}$$

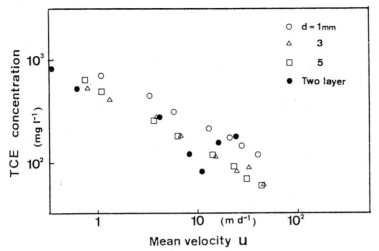

Fig. 5 Dissolution of TCE in water. Black circle indicates the result in the water-undiluted TCE two-layer system, and other symbols indicate the difference of the glass beads in the porous media two-layer system.

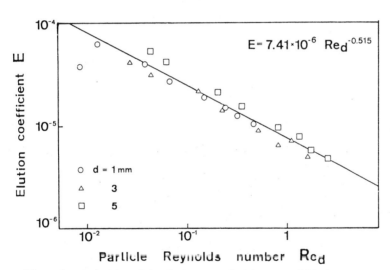

Fig. 6 Relationship between elution coefficient E and particle Reynolds number

Monitoring of groundwater pollution by gaseous contents in surface soil

The organochlorines are volatile, therefore their gases reach the surface area when the subsurface environment is contaminated by the organochlorines. In this technique, some sorbents are utilized to trap the contaminants which diffuse through the soil (Marrin and Kerfoot, 1988). Two types of the soil gas monitoring method were applied to a model region of groundwater pollution due to tetrachloroethylene which seems to come from the waste landfill site.

Soil gas survey by fingerprint method

The volatiles are collected on activated charcoal glued to a wire placed at 30cm depth below the ground surface for a couple of weeks, and are analysed with the pyrolysis mass-detector (Fingerprint method by NERI-Petrex). The model region is located in Nagano Pref. in Japan. The field is in a mountaineous area, and six water springs are observed in the southern part of the field. These springs are famous from old times for their clean drinkable water supplied for regional town people, but several years ago tetrachloroethylene (PCE for short) was detected, and the concentration overshot the provisional standards for drinking water in Japan (0.01mg/l). We had no clue to find out the pollutant source because there is no surface water around there. Then, the contaminants in surface soil gas were examined by detectors of this method at about 100 sites in this area at 100 m intervals. Fig. 7 illustrates the ion count contours with respect to PCE, and the zone with high value stretching from the waste landfill site in the northern part to the springs can be recognized. On the basis of this monitoring, the waste landfill site is firmly specified to be the pollutant source of the contaminated spring water.

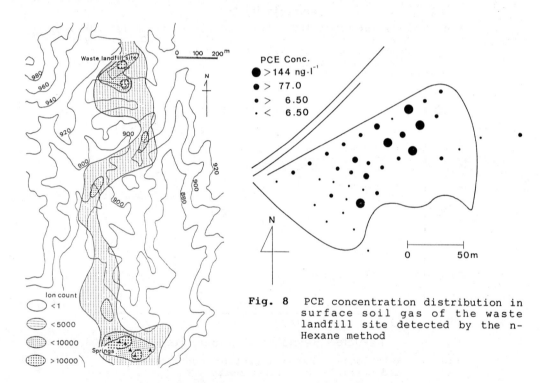

Fig. 8 PCE concentration distribution in surface soil gas of the waste landfill site detected by the n-Hexane method

Fig. 7 PCE ion-count distribution in surface soil gas detected by fingerprint method.

Soil gas survey by n-Hexane method

For shortening the investigation time, a new tool to trap the soil gas from the volatile contaminants was devised at the National Institute for Environmental Studies, Japan. This technique is that first a small hole of 2cm in diameter and 30cm in depth is dug, then soil gas of 20 ml is sampled with a gas-tight syringe from the bottom of the hole and the volatile contaminants are extracted in n-Hexane placed in a vial at the field. A process for taking one sample can be completed within fifteen minutes. The volumes of the vial and n-Hexane are 1 ml and 0.8 ml, respectively, and at the laboratory the contaminants are analyzed by gas chromatography with an electron capture detector.

The field survey using the n-Hexane method was done at forty points at the waste landfill site as above mentioned. The result of analysis of PCE concentration in the surface soil gas at the landfill site is shown in Fig. 8.

Summary

The basic experiments to solve the mechanisms of groundwater pollution are still important for framing the works for improvement, conservation, control of groundwater quality and subsurface environment. In order to specify the polluted area and pollutant source, the soil gas monitoring is very effective to understand the circumstances of groundwater pollution in early stages. Though the concentration of the volatile contaminants in the surface soil gas is highly dependent on the geological condition, the soil gas surveying is the technique to promote broadly as an available monitoring method.

APPENDIX : DIRECTIONS BY NATIONAL GOVERNMENT, JAPAN

After the groundwater pollution by TCEs was well known by people in the country, the ministries concerned have had various committees to discuss the countermeasure and directed some guidance to the local governments, as follows:

1984 : MHW* notified the temporary standards of TCEs for drinking water and informed the manual of TCEs analysis techniques.
1984 : EA* showed the guideline of target standards of TCEs for both the infiltration in the subsurface and the effluent to public water area.
1984 : MITI* gave notice of standards of TCEs effluent concentration from industries.
1984 : MC* directed TCEs effluent concentration to sewerage system.
1984 : MHW made the manual for appropriate usage of PCE at dry-cleaning shops and informed the target concentration of effluent from the treatment equipment of waste water.
1984 : MHW notified the appropriate control of the wastes including TCEs.
1985 : MITI issued the manual describing the appropriate usage and control of TCEs in industrial process.
1986 : MHW informed the necessity of appropriate control and routine checkup of the quality of drinking-well-water, and directed the measures on accident.
1986 : EA issued a guideline as to the field survey and observation of groundwater pollution area.
1989 : EA made a partial amendment of The Law of Water Pollution Control in order to prevent groundwater from pollution and to take measures for accident.

```
*note     EA : Environment Agency of Japan
         MHW : Ministry of Health and Welfare
        MITI : Ministry of International Trade and Industry
          MC : Ministry of Construction
```

ACKNOWLEDGEMENT

A part of this study is supported by TORAY Science and Technology Grants. The authors wish to thank the persons concerned of Mitsui Mineral Development Engineering Co., Ltd. for their cooperation of field survey.

REFERENCES

Hirata, T. and K. Muraoka(1988): Vertical migration of chlorinated organic compounds in porous media, Water Research, 22(4), 481-484.
Japan Environment Agency(1983): Field data of groundwater contamination in 1982 (in Japanese).
Marrin, D.L. and H.B. Kerfoot(1988): Soil-gas surveying techniques, Environ. Sci.Tecnol.,22(7),740-745.

FATE AND TRANSPORT OF ALACHLOR, METOLACHLOR AND ATRAZINE IN LARGE COLUMNS

B. J. Alhajjar, G. V. Simsiman and G. Chesters

Water Resources Center, University of Wisconsin, 1975 Willow Drive, Madison, WI 53706, USA

ABSTRACT

^{14}C ring-labelled atrazine, alachlor, and metolachlor were surface-applied at 3.14 kg a.i./ha in greenhouse lysimeters containing two soils in an ongoing experiment. Bromide (Br) -- a conservative tracer -- at 6.93 kg/ha as KBr and nitrate-nitrogen (NO_3-N) at 112 kg/ha as KNO_3 were mixed with each herbicide and surface-applied. Growth of Red top (*Agrostis alba*) was established in each column (105 cm long and 29.4 cm i.d.). The experiment consisted of 12 columns (2 soils x 3 herbicides x 2 replicates) each fitted with four sampling ports for leachates, a volatilization chamber, and an aeration and irrigation system. Volatile materials are being trapped directly in solvents. One column replicate was dismantled for soil and plant analyses. Columns of Plainfield sand and Plano silt loam treated with alachlor and metolachlor were sampled after 23 and 28 weeks, respectively; the atrazine columns after 35 weeks. Herbicide residues are determined by liquid scintillation counting, extracted and separated by thin-layer chromatography using autoradiographic detection. Volatilization was $\leq$ 0.01% of the amount of herbicide applied. The order of herbicide mobility was alachlor > metolachlor >> atrazine. As many as 8 to 12 alachlor metabolites and 2 to 6 metolachlor metabolites were separated in leachates.

KEYWORDS

Alachlor; metolachlor; atrazine; herbicides; soil columns; leaching; volatilization; nonpoint pollution; herbicide mass-balance.

INTRODUCTION

The United States Environmental Protection Agency (USEPA) provide criteria for identifying and evaluating the source and spread of nonpoint pollution. Nonpoint pollution is the movement of pollutants from diffuse sources into surface or subsurface waterbodies and is among the leading causes of water quality problems (Chesters and Schierow, 1985). Pesticides are among the most important nonpoint pollutants carried from agricultural fields to surface waters or infiltrated into groundwaters. In the last decade, the occurrence of pesticides in ground water has been confirmed (Cohen et al., 1984; Cohen et al., 1986; USEPA, 1986). The USEPA's Office of Ground Water Protection compiled a list of 17 pesticides in 23 states found in groundwater due to normal land applications (USEPA, 1986). Pesticides in groundwater cause public concern about the quality of drinking water, which has led to renewed pesticide monitoring and research and the call for stricter health advisories and water quality standards (Ehart et al., 1986). Pesticides may also escape to the atmosphere by volatilization (Taylor, 1978; Jury et al., 1980).

In several regions of Wisconsin, evidence shows the presence of a variety of herbicides in groundwater including atrazine [2-chloro-4-(ethylamino)-6-(isopropylamino)-s-triazine], metolachlor [2-chloro-N-[2-ethyl-6-methylphenyl]-N-2(2-methoxy-1-methyl) acetamide], and alachlor [2-chloro-2',6'-diethyl-N-(methoxymethyl) acetanilide]. They are extensively used in corn, potato and soybean production (WDATCP, 1986). A groundwater pesticide monitoring program in Wisconsin has shown detectable amounts of alachlor in 75 of 830 wells sampled, of metolachlor in 59 of 510 wells, and of atrazine in 152 of 752 wells (WDNR, 1988). Under normal agricultural conditions, 4 of 14 wells were contaminated with alachlor, 4 of 10 with metolachlor and 5 of 13 with atrazine (Postle, 1987). Groundwater contamination by these herbicides has led to increased public pressure to regulate their use because of potential adverse effects on human health; alachlor is listed as a probable human carcinogen while metolachlor and atrazine have shown indications of carcinogenicity to rats. Recently, groundwater standards for alachlor, atrazine, and metolachlor were set for Wisconsin at 0.5, 3.5, and 15 μg/l (ppb), respectively.

Alachlor, metolachlor and atrazine are now under close scrutiny by the state of Wisconsin and federal agencies with regard to their environmental fate and toxicological significance to humans, livestock, and ecosystems. The objective of this ongoing investigation is to evaluate transport of the herbicides and include volatilization, sorption, plant uptake, and degradation in a mass balance.

MATERIALS AND METHODS

Soil Types

Two Wisconsin soil types were used: Plainfield sand (Typic Udipsamment) obtained from a farm in Plover, WI; and Plano silt loam (Typic Argiudoll) obtained from the Rock County Experimental Farm near Janesville, WI. Pits (1 m^2) were dug for each soil to a depth of 1 m. Each soil was separated by horizons. Five horizons were recognized for the Plainfield soil and four for the Plano soil. The soils were air-dried, passed through a 6 mm screen and used for column packing after each horizon was well mixed; large rocks and roots were removed from the soils. Representative samples from each soil horizon were analyzed (Table 1).

TABLE 1 Description of Surface and Subsurface Soils

Property	Plainfield sand, depth (cm)					Plano silt loam, depth (cm)			
	0-18	18-25	25-58	58-79	79-100	0-25	25-36	36-51	51-100
pH	5.8	6.0	5.7	5.8	5.9	6.5	6.4	5.9	5.6
Organic matter, %	2.0	0.9	0.3	0.3	<0.1	3.5	1.5	1.0	0.6
Total Kjeldahl N, μg/g	690	390	220	160	90	1,800	920	590	440
Available P, μg/g	79	45	55	82	64	38	15	24	48
Ca, μg/g	350	340	160	40	50	1,700	1,500	1,700	1,900
Mg, μg/g	80	70	50	20	10	630	640	880	870
K, μg/g	61	31	26	11	7.0	210	140	190	200
Na, μg/g	4.0	3.0	2.0	2.0	2.0	13	19	19	20
NH_4-N, μg/g	<0.5	<0.5	<0.5	<0.5	<0.5	6.5	4.0	2.0	<0.5
NO_3-N, μg/g	0.5	0.5	0.5	0.5	0.5	7.5	<0.5	<0.5	<0.5
Cl, μg/g	<0.5	<0.5	<0.5	<0.5	<0.5	2.5	1.5	1.0	0.5
Br, μg/g	<0.1	<0.1	<0.1	<0.1	<0.1	<0.1	<0.1	<0.1	<0.1
CEC,[a] meq/100 g	2.6	2.4	1.3	0.4	0.3	14	13	16	17
Sand, %	91	89	90	93	97	9	12	10	7
Silt, %	6	6	5	4	1	70	62	57	62
Clay, %	3	5	5	3	2	21	26	33	31

[a]Cation exchange capacity estimated from cation composition of soil.

Soil Columns

The experimental set-up shown in Figure 1 consists of twelve polypropylene columns (105 cm long x 29.4 cm internal diameter with 1.7 cm wall thickness). Controversy exists about the best way to construct columns for transport experiments and so details are provided of the materials and dimensions used. Pure polypropylene is a rigid material slightly less inert than teflon but much less expensive; and far more inert to sorption than polyethylene, polyvinyl chloride, and silicone rubber (Barcelona et al., 1985).

Pure polypropylene plates (35.5 cm x 35.5 cm with 1.7 cm wall thickness) were cut and securely attached to the bottom of each column using stainless steel screws mounted into holes die-threaded into the plates and the edges of the columns. The plates were tightly sealed to the

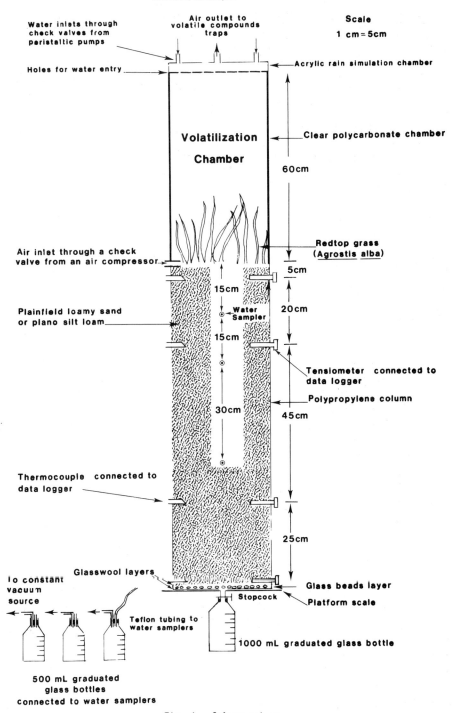

Fig. 1. Column set-up

columns by placing a teflon ring cut to form a gasket seal along the walls of the columns and plates. Tap-threaded pure polypropylene adapter tubes of 3.2 mm internal diameter were attached to a hole die-threaded into the center of each plate. The adapter tubes were connected to stopcocks to control the free flow of leachate. Teflon thread seal ribbon was used to seal the tap- and die-thread combinations. To further insure air-tight conditions, hot glue was applied with a heat gun to all sealing surfaces. A hot air stream from a coil-heated air gun was used to form a smooth layer of plastic glue that hardens in 1 minute. A layer of fiber glass wool was placed at the bottom of each column above the adapter tube to prevent soil being washed from the columns. Solid glass beads (4 mm in diameter) were layered at a depth of 1 cm above the bottom of each column to create a saturated boundary. Another layer of fiber glass wool was placed on top of the glass beads to separate the beads from the soil. The soil horizons were carefully packed to their original field bulk density.

To draw water samples from several depths of the soil columns, vacuum-pressure sampling systems were designed and tested. Each sampling system was constructed by attaching a fritted (porous) glass disc in the form of a tube to a vacuum pump with an adjusted pressure of 0.1 bar. The diameter of the fritted disc was 3 cm with nominal pore size of 10 to 15 μm and 18 cm long with a 0.8 cm internal diameter tube reinforced by passing through rigid teflon tubing (1 cm I.D. and 2 mm wall thickness) to withstand the pressures from soil packing. A fritted glass lysimeter was developed and tested for soil water sampling because glass is known for its relative inertness compared to other materials. Water samplers were placed at 15, 30, and 60 cm below the soil surface. Prior to packing the soil, holes of 13 mm in diameter were drilled to mount the samplers. Care was taken to securely mount each glass sampler tube encased in rigid teflon tubing to the column wall with teflon film and hot glue to ensure complete sealing.

To provide for evapotranspiration, Redtop (*Agrostis alba*) grass was grown on the columns. Redtop is a long-lived perennial grass growing to a height of up to 1 m. Its root system consists of shallow, vigorous root stalks--5 to 15 cm long--that form a loose, coarse turf. It is one of the best wetland tame grasses. It can remain underwater for considerable periods of time without damage, yet it adapts to dry conditions and acid or alkaline soils. It is used as a soil binder along waterways.

Each herbicide was evenly sprayed on the soil surface with irrigation water at 3.14 kg a.i./ha (21.3 mg herbicide/column) after establishment of the grass cover. ^{14}C-labelled herbicides with the highest specific activities available were used; 72.65, 79.3 and 103.2 μCi/mg for alachlor, metolachlor, and atrazine, respectively. Application solutions consisted of 32, 30 and 23% ^{14}C labelled alachlor, metolachlor, and atrazine, respectively. Application amounts were at field recommended rates. Bromide (Br) -- a conservative tracer -- was applied to each column at 6.93 kg/ha as KBr. The soils were fertilized with NO_3-N at 112 kg/ha as KNO_3.

The columns were fed intermittently with distilled water to simulate rainfall. Distilled water was applied according to a precipitation schedule based on rainfall averages from 30 year records for Hancock and Janesville, WI compiled from National Oceanic and Atmospheric Administration records at the National Climatic Data Center, Ashville, North Carolina. Water was added to each column at regular intervals. The amount of water added accommodated the rainfall and sampling volumes which was an amount similar to the irrigation at the Hancock Experimental Farm.

Temperature and water potential profiles for the columns were determined using thermocouple junctions and tensiometers with pressure transducers. Four columns representing duplicates of the two soil types were monitored for temperature and water potential changes at four soil depths: 5, 25, 70, and 95 cm below the soil surface. Another thermocouple was placed at the soil surface-atmosphere interfaces. The thermocouples were teflon insulated. Each tensiometer was designed and constructed from a fritted glass disc, attached to a glass tube and interfaced with a pressure transducer. The fritted disc had a diameter of 3.0 cm and nominal porosity of 4.0 to 5.5 μm. Such a disc was tested and found to withstand very low tensions. Porous ceramic cups are not recommended for use because they sorb large amounts of organic compounds. The thermocouples and pressure transducers were attached to a Model 21X data logger (Campbell Scientific Inc., Logan, UT).

Evapotranspiration measurements or water loss from the soil surface and grass for four columns representing duplicates of the two soil types were monitored closely by placing these columns on scales (Fairbanks Scales, St. Johnsbury, VT) serving as continuous weight control devices. Evapotransporation measurements were taken prior to each scheduled precipitation by weighing the columns. Evapotranspiration was determined by keeping records of irrigation inputs, drainage and change in water storage.

Clear polycarbonate cylinders (0.16 cm wall, 30.5 cm I.D. and 60 cm long) were placed on top of each column, joined by a teflon ring, and sealed air-tight with hot-melt glue. Rain simulator chambers were constructed by sealing two plexiglass plates (0.34 cm wall x 36.5 cm x 36.5 cm) to sandwich a plexiglass cylinder (0.32 cm wall, 30.5 cm I.D., and 5 cm long). Holes

(19 gauge) were evenly made in the bottom plate. Two pairs of plexiglass tubes (0.32 cm I.D. and 5 cm long) were installed in the upper plate to provide an inlet for water and outlet for air, respectively. The water inlets distribute water evenly across the area into the rain simulator chambers. The air outlet tubes connect the outside of the column to the rain simulator, and volatilization chambers to allow free entry of water between the volatilization chamber and the soil surface. The rain simulator chambers were sealed to the polycarbonate cylinders of each column with acrylic glue. Water is delivered to the chamber through plexiglass tubes passing by check valves using Masterflex multichannel peristaltic pumps controlled by a programmable electronic timer. The air outlets are merged by a T-shaped glass connector attached to a check valve and through a series of a back trap and two solvent traps containing a 1:1 mixture of hexane:acetone and another back trap and two $^{14}CO_2$ traps (Simsiman and Chesters, 1975; Marinucci and Bartha, 1979). The traps are changed weekly.

The atmosphere contained in the volatilization chamber was changed four times a day by passing fresh air through a check valve sealed to a hole made into the polycarbonate cylinder 5 cm above the polypropylene column. Air was delivered from an oilless air compressor. Every 6 hours the pump was turned on and off by a programmable Chron Trol electronic timer (Lindburg Enterprises, San Diego, CA). The air flow was measured by a Gilmont air flowmeter. It was calculated that total air volume of the volatilization chamber is exchanged by running the pump for 5 minutes.

Sampling and Analyses

Water samples from the three soil depths were drawn at week 1, 2, 3, 4, 6, 8, 11, 15, and 23 by applying a low vacuum to the fritted glass lysimeter at each sampling port. The samples were collected in 500 ml glass bottles. The leachate from the bottom of each column is sampled at regular intervals in 1,000 ml glass bottles through stopcocks. The glass bottles were all new, graduated and fitted with teflon caps. They were washed three times with methanol before sampling. After recording the volume, herbicides were extracted from water samples by extraction with ethyl acetate immediately after sampling. The 1:1 hexane:acetone mixture air traps and the $^{14}CO_2$ traps were sampled weekly. Water and volatile materials were quantified by liquid scintillation counting (LSC) and all samples were refrigerated at 4°C.

One replicate of each herbicide treatment was dismantled and the other remained intact for later evaluation. Columns were dismantled removing volatilization and irrigation chambers. Shoots and roots of the grass were harvested separately. Columns were drained by gravity before soil sampling and the last leachate volume was extracted with ethyl acetate and evaluated by LSC. Samples of plant roots and shoots were collected from each column for herbicide residue analysis of plant tissue. Soil samples from four layers of each column were collected manually at 0-15, 15-30, 30-60, and 60-100 cm. In the process, every layer was sampled in three locations using borehole sampling with a long-reach spatula at the center of the column, near column walls, and between the center and the walls of columns. Each layer was mixed thoroughly and a fourth sample was taken from the mixed (bulk) layer. Plant tissue samples were carefully separated from soil particles by thoroughly washing with distilled water, and moisture contents of soil and plant tissue were measured. Total $^{14}CO_2$ in soil and plant samples were determined by combustion using biological oxidizer model OX600 (R. J. Harvey Instrument Corp., Hillsdale, NJ). The soil and water samples were also extracted with ethyl acetate for analysis by thin layer chromatography (TLC).

Herbicide Residue Extraction, Separation and Identification

Aliquots of 100 ml of water samples were extracted twice with ethyl acetate (100 ml). The extracts were dried by filtration through anhydrous sodium sulfate (Na_2SO_4) and evaporated to dryness in a stream of air at 45°C in a Rotovac rotary evaporator. The residues were redissolved in 2 ml ethyl acetate and stored in the freezer.

Extraction of herbicide residues was accomplished by adding 50 ml of ethyl acetate to approximately 50 g of herbicide-treated soil in a wide-mouth 240 ml glass bottle fitted with a teflon cap. The soil samples were shaken overnight, filtered through a Buchner funnel fitted with a 9 cm diameter Whatman glass microfiber filter 934-AH and 20 g oven-dried sodium sulfate. The samples were reextracted two times with 50 ml portions of ethyl acetate, shaken for 1 hr and filtered each time through the Buchner funnel. The combined filtrate, evaporated to dryness using a Rotovac rotary evaporator, was redissolved in 2 ml of ethyl acetate and stored in a freezer.

The herbicide residues in ethyl acetate extracts from water and soil samples were separated and analyzed by TLC using autoradiographic detection. Aliquots (200 μl) of the extracts were spotted on commercial silica gel plates and developed with the proper solvent mixtures for each herbicide. The solvent mixtures used were benzene:methanol (19:1) for alachlor, hexane:chloroform:ethyl acetate (3:1:1) for metolachlor, and benzene:acetic acid:water (60:40:3) for atrazine. Radioactive areas on the TLC plates were detected by placing Kodak SB-5 X-ray films on the plates. The radioactive spots representing the herbicide residues were scraped from the plates and transferred in 2 ml high purity methanol into teflon capped 8 ml vials. Aliquots (0.1 ml) were transferred to scintillation vials and radioactivity counted. Alternatively, radioactivity was quantified using a TLC-scanner.

RESULTS AND DISCUSSION

The cumulative amount of herbicide volatilized (organic volatile residues and $^{14}CO_2$) from soil columns is shown for 11 weeks in Table 2. The amounts were small possibly because the herbicides were applied with irrigation water and quickly leached below the soil surface. $^{14}CO_2$ also may have been dissolved by soil water and transported deeper into the profile. Volatilization may have been unusually low because conditions were essentially stagnant and the effect of wind was ignored. $^{14}CO_2$ was in general slightly greater than other trapped ^{14}C volatiles. Volatilization will continue to be monitored but it is unlikely to produce further useful information.

TABLE 2 Cumulative ^{14}C Volatilization of Herbicide Residues and CO_2 from Soil Columns in 11 weeks[a]

Soil	^{14}C (DPM)			^{14}C (%)		
	Volatiles	CO_2	Total	Volatiles	CO_2	Total
			Alachlor			
Plainfield sand	5×10^3	9×10^3	14×10^3	4.1×10^{-4}	8.2×10^{-4}	1.2×10^{-3}
Plano silt loam	26×10^3	24×10^3	50×10^3	2.3×10^{-3}	2.2×10^{-3}	4.5×10^{-3}
			Metolachlor			
Plainfield sand	24×10^3	92×10^3	116×10^3	2.2×10^{-3}	8.2×10^{-3}	1.0×10^{-2}
Plano silt loam	6×10^3	5×10^3	11×10^3	5.5×10^{-4}	4.3×10^{-4}	9.8×10^{-4}
			Atrazine			
Plainfield sand	13×10^3	78×10^3	91×10^3	1.2×10^{-3}	6.9×10^{-3}	8.1×10^{-3}
Plano silt loam	8×10^3	115×10^3	123×10^3	7.1×10^{-4}	1.0×10^{-2}	1.1×10^{-2}

[a]Values are means from two columns for each herbicide in disintegrations per minute (DPM) and percentage of total herbicide applied. Initial ^{14}C added was 1.0993×10^9, 1.1249×10^9, and 1.1224×10^9 DPM as alachlor, metolachlor, and atrazine, respectively.

Amounts of applied ^{14}C herbicide recovered from soil and leachate are recorded as total herbicide residues (Table 3). The order of herbicide mobility is alachlor > metolachlor >> atrazine. More than twice as much alachlor as metolachlor leached to the bottom of the

TABLE 3 Mass Balance of ^{14}C in Columns

Herbicide	Time (Weeks)	Leachate ^{14}C (%) at Depth (cm)				^{14}C Recovered from Soil by Combustion (%)	Total Recovery (%)
		15	30	60	100		
				Plainfield Sand			
Alachlor	23	0.21	-	4.9	17	76	98
Metolachlor	23	0.32	0.12	1.2	7.3	95	104
Atrazine	35	1.4	0.22	0.04	0.12	102	104
				Plano Silt Loam			
Alachlor	28	-	0.88	1.0	4.9	64	71
Metolachlor	28	1.5	1.0	0.47	0.92	83	87
Atrazine	35	1.8	0.32	0.13	0.08	72	74

- No sample.

Plainfield sand columns while five times as much alachlor as metolachlor reached 1 m in the Plano soil columns. ^{14}C atrazine levels were low in the leachates from both soils. Continuation of the leaching experiments using the remaining columns will provide further information with respect to the mobility of herbicide residues. Based on determination of ^{14}C

in leachates and $^{14}CO_2$ release from soil samples by combustion, a quantitative mass balance can be achieved for the three herbicides in Plainfield sand. However, in Plano silt loam only 71-87% of the ^{14}C can be accounted for (Table 4). This suggests that the herbicides and/or their degradation products are trapped in the interlayer spaces of expanding clay minerals in the Plano silt loam. The evidence is preliminary and the dimensions of the interlayer spacings need further investigation.

Radiolabelled herbicide distributions in the profiles of the two soils are shown in Table 4. The three herbicides and/or their metabolites were concentrated in the top layer (0-15 cm) of both soils and levels decreased with increasing depths in the soil profile.

TABLE 4 Distribution of ^{14}C in Columns

% Alachlor at Depth (cm)				% Metolachlor at Depth (cm)				% Atrazine at Depth (cm)			
0-15	15-30	30-60	60-100	0-15	15-30	30-60	60-100	0-15	15-30	30-60	60-100
				Plainfield Sand							
43	15	9.1	8.8	46	21	15	13	85	9.3	3.9	3.3
				Plano Silt Loam							
33	15	8.9	6.7	49	18	10	5.9	48	13	7.5	3.6

Herbicide residues were detected by TLC in extracts of leachate from the bottom ports of the six sacrificed columns. Atrazine did not leach in amounts large enough to allow extraction of quantities measurable by TLC except at the upper sampling port. Detectable amounts of alachlor and metolachlor and their metabolites moved completely through the columns. In Plainfield sand, alachlor was not detected before week 16; it appeared at and after week 16. In Plano silt loam, alachlor was not detected at or before week 16, was detected at week 17 but not at week 19. Metolachlor was detected in leachate from Plano silt loam at and after week 17, and at and after week 19 in Plainfield sand.

For the Plainfield sand, alachlor showed as many as 12 metabolites leaching from the bottom of the column after 23 weeks of which 4 were judged to be present in major amounts; for metolachlor it was 6 metabolites with 2 judged to be major. For the Plano silt loam alachlor displayed 8 (2 of them major) metabolites and metolachlor showed 2 (1 major) metabolite. There was insufficient leaching of atrazine residues to make these types of separations. TLC indicates that some metabolites of alachlor and metolachlor were more mobile than the parent compounds. The mobility of alachlor was slower in Plano silt loam than in Plainfield sand, but alachlor seems to be degraded faster in Plano silt loam than in Plainfield sand.

Separation of herbicide metabolites from water and soil samples will continue by TLC and HPLC. Identification of major metabolites of alachlor, metolachlor, and atrazine is being attempted using separation, purification and MS techniques. A data base is being constructed for movement of the three herbicides to the 1 m depth with time. Attempts will be made to extrapolate the data to greater soil depths. The data base will be used to calibrate and verify such available pollutant movement models as the Leaching Estimation and Chemistry Model (Wagenet and Hutson, 1987) and the U.S. EPA Pesticide Root Zone Model (Carsel et al., 1984).

REFERENCES

Barcelona, M.J., Helfrich, J.A. and Garske, E.E. (1985). Sampling tubing effects on groundwater samples. Anal. Chem., 57, 460-464.
Carsel, R.F., Smith, C.N., Mulkey, L.A., Dean, J.D. and Jowise, P. (1984). Users manual for the pesticide root zone model (PRZM). EPA-600/3-84-109. U.S. Environmental Protection Agency, Athens, GA, 216 pp.
Chesters, G. and Schierow, L. (1985). A primer on nonpoint pollution. J. Soil Water Conserv., 40, 9-13.
Cohen, S.Z., Creeger, S.M., Carsel, R.F. and Enfield, C.G. (1984). Potential pesticide contamination of groundwater from agricultural uses. In: Treatment and Disposal of Pesticide Wastes, R.F. Krueger and J.N. Seiber (Eds). ACS Symp. Series 259. American Chemical Society, Washington, D.C., pp. 297-325.
Cohen, S.Z., Eiden, C. and Lorber, M.N. (1986). Monitoring ground water for pesticides. In: Evaluation of Pesticides in Ground Water, W.Y. Garner, R.C. Honeycutt and H.M. Nigg (Eds). ACS Symp. Series 315. American Chemical Society, Washington, D.C., pp. 170-196.
Ehart, O.R., Chesters, G. and Sherman, K.J. (1986). Ground water regulations: Impact, public acceptance, and enforcement. In: Evaluation of Pesticides in Ground Water, W.Y. Garner, R.C. Honeycutt and H.N. Nigg (Eds). ACS Symp. Series 315. American Chemical Society, Washington, D.C., pp. 488-498.

Jury, W.A., Grover, R., Spencer, W.F. and Farmer, W.J. (1980). Modeling vapor losses of soil-incorporated triallate. Soil Sci. Soc. Am. J., 44, 445-450.

Marinucci, A.C. and Bartha, R. (1979). Apparatus for monitoring the mineralization of volatile ^{14}C-labeled compounds. Appl. Environ. Microbiol., 38, 1020-1022.

Postle, J. (1987). Groundwater monitoring project for pesticides. Progress Report submitted by Wisconsin Department of Agriculture, Trade and Consumer Protection to Wisconsin Department of Natural Resources, Madison, WI, 13 pp.

Simsiman, G.V. and Chesters, G. (1975). Persistence of diquat in the aquatic environment. Water Research, 10, 105-112.

Taylor, A.W. (1978). Post-application volatilization of pesticides under field conditions. J. Air Pollut. Control Assoc., 28, 922-927.

U.S. Environmental Protection Agency (USEPA). (1986). Pesticides in ground water: Background document. Office of Groundwater Protection WH5506, Washington, D.C.

Wagenet, R.J. and Hutson, J.L. (1987). Leaching estimation and chemistry model (LEACHM). Water Resources Institute, Center for Environmental Research, Cornell University, Ithaca, NY, 80 pp.

Wisconsin Department of Agriculture, Trade and Consumer Protection (WDATCP). (1986). Wisconsin 1985 pesticide use. Wisconsin Department of Agriculture, Trade and Consumer Protection to Wisconsin Department of Natural Resources, Madison, WI, 13 pp.

Wisconsin Department of Natural Resources (WDNR). (1988). Summary of groundwater pesticide monitoring for 07/01/83 thru 06/30/87. Prepared by staff of WDNR, June 23, 1988, 3 pp.

MINERAL SAND MINING AND ITS EFFECT ON GROUNDWATER QUALITY

M. N. Viswanathan

Water Investigation and Planning Section, Hunter Water Board, Newcastle West, Australia 2290

ABSTRACT

Tomago sandbeds (New South Wales, Australia) is a coastal unconfined aquifer, where mining of mineral sands, like Rutile, Zircon, Ilmenite etc., was in progress since 1972. Groundwater is also extracted from Tomago aquifer for urban water use. Groundwater iron levels vary from 0.1 mg/litre to about 10 mg/litre. Iron in excess of 0.3 mg/litre is removed by chemical treatment. Mining of mineral sands resulted in the substantial increase of iron levels. The level of increase itself being very site specific. Several processes were identified as being responsible for such increases. If water were to be extracted from the mined area, additional treatment would be required to remove excess iron.

KEYWORDS

Groundwater, mining, mineral sands, water quality, iron, sulphate, aquifer, water treatment.

INTRODUCTION

Tomago sandbeds is an unconfined coastal aquifer located 20 km north of Newcastle, Australia. The total catchment area is about 130 km^2. Tomago is an important source of water for Newcastle region. About 30% of urban water supply comes from Newcastle. The aquifer consists of fine to medium sand with occasional traces of clay lenses. The depth of the aquifer is about 20 m. and has an average transmissivity of about 400 m^2/day and a specific yield of about 0.15. There are about 200 production bores and extraction of water from the aquifer varies between 10 - 70 ML/day depending on the demand. The groundwater quality is very site specific. The average groundwater quality being: total iron - 5.0 mg/Litre, Chloride - 40 mg/Litre, Sulphate - 35 mg/Litre and pH 4.5.

The Tomago aquifer contains minerals like Rutile (Oxide of Titanium), Zircon(Zirconium Silicate), Ilmenite (Oxide of Titanium and Iron), Monazite (Phosphate of Rare Earths), etc. The mineral sand mining has been in progress at Tomago since 1972. Mining is carried out in an artificially constructed pond about 120 m square with a water depth of about 5 m on which is floated a suction cutter dredge that moves across the front of the pond using a water jet to entrain the soil from the ground surface to the bottom of the mineralised zone. The entrained soil slurry is pumped to a concentrating plant, which floats behind the dredge and which separates the sand from the

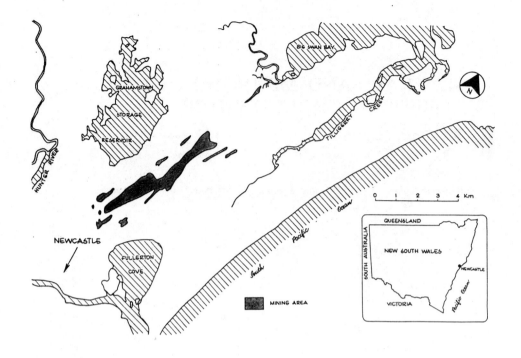

Figure 1. Tomago sandbeds and mined area

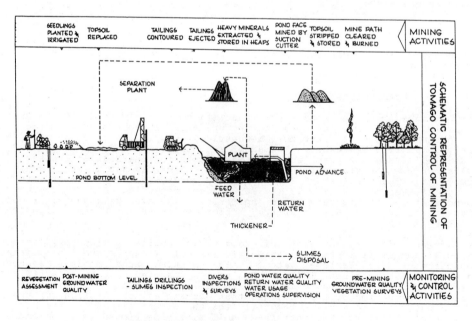

Figure 2. Mining process

heavy mineral. The rejected sand is pumped as a high density slurry to the rear of the pond as tailings. The water used for processing, entrains a quantity of fines or 'slime' which is removed by settling in a thickener and the recovered water returned to the pond and reused in the dredging operation. The removed slime is disposed of outside the catchment area of the sandbeds. The mining operation is shown in figure 2.

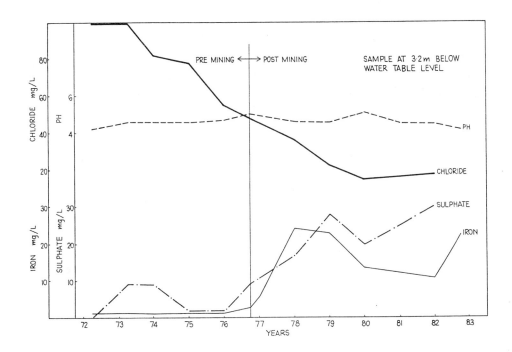

Figure 3. Effect of mining on water quality

GROUNDWATER QUALITY VARIATIONS

Mining of minerals, coal etc., within an aquifer usually affects groundwater quality. For example, extraction of coal has resulted in increased iron values of groundwater (Corbett 1977, Emrich 1969). Groundwater pollution resulting from leaching of old mine tailings was reported by Mink et al (1972). Dispersion of heavy metals has occurred either through the physical movement of tailings material into creeks or by leaching of groundwater through and at the base of the dumps. The tailings dumps were accumulated at the ground surface due to lead and copper mining in Western Australia (Mann et al 1983). Brassington (1982) concludes that when groundwater is exposed in quarries or mine workings it is susceptible to changes in its quality by various physicochemical or biochemical processes that are not active within the undisturbed aquifers, and cites several examples where groundwater was affected by various mining operations.

WATER QUALITY MONITORING

Groundwater quality monitoring was undertaken ever since mining began at Tomago. Variations in water quality were investigated by testing water samples from a network of observation wells. Since water quality is very site specific at Tomago, observation wells were established at a spacing of about 300 m. These wells were installed about 4 - 5 years prior to mining and water samples were collected from these wells. These were analysed for total iron, sulphate, chloride, pH etc., After the area was mined, these observation wells were re-established at the same locations and monitoring continued for a period of about five years after mining.

Figure 3 shows variation in water quality before and after mining. The drop in chloride between 1972 and 1980 probably represents increased infiltration due to heavy rainfall during these years. Since 1980 the rainfall levels were low and hence a slight increase in chloride levels. The average rainfall for Tomago is about 1100 mm per year. Between 1972 and 1980 rainfall levels were well above the average levels. Since 1980 rainfall levels varied between 600 and 1000 mm per year. No trend in the variation of pH was observed before and after mining. Significant increases in sulphate and iron levels were observed after mining. Figure 4 shows variations in sulphate level at various depths of aquifer. Approximately three levels were chosen for analysis. These being, one at mining pond top, which is near the water table level, and the second at middle of pond depth which is about 3 m depth from water table level and the third at pond bottom level, which is approximately about 3.5 - 5 m below the water table level. Increases in sulphate levels do not occur at the same time at these three levels after mining. This is even more evident in increases in iron levels as shown in figure 5. From urban water use point of view, increases in iron levels are more important than any other parameter. If water were to be extracted from the mined area for urban water use, increased water treatment would be required to remove the increased iron levels.

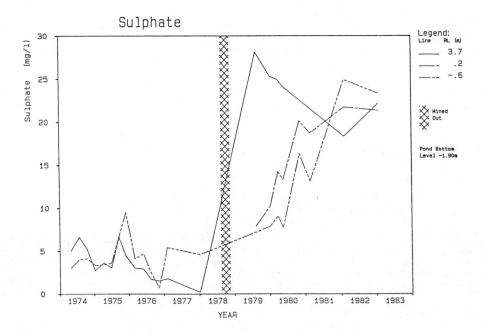

Figure 4. Variation of sulphates with depth

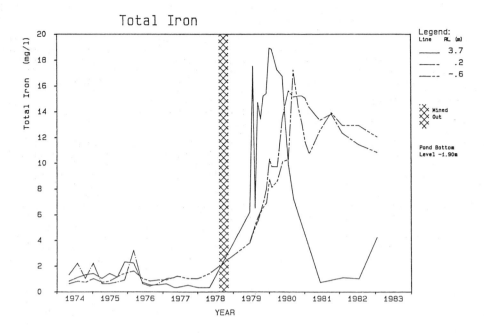

Figure 5. Variation of total iron with depth

This would add to the cost of operations of the water authority responsible for water supply. Iron levels for urban water use is limited 0.3 mg/Litre.

The mining process, as described previously, disturbs the equilibrium conditions within the aquifer. The impact of mining process on the aquifer could be due to:

i.   Mixing of sand (dredging, separation, and redeposition).

ii. Stripping of sand at the time of minerals separation.

iii. Change in groundwater temperature (due to open mine pond).

iv. Aeration of groundwater (within the mine pond), and

v.   Exposure to sunlight (within the mine pond).

Mixing of sand results in taking biologically rich top soil to lower depths of aquifer. In addition, organic nutrient material is also transported from top soil to lower depths. Such an action could increase the biological activity at lower depths, particularly in the growth of anaerobic bacteria like Bacillus circulans etc, which are capable of reducing iron and bringing it into solution.

Stripping of sand, at the time of minerals separation, increases surface area, thereby increasing the chemical activity.

Change in groundwater temperature, particularly an increase in temperature, could accelerate chemical activity. Groundwater temperature at Tomago is usually about $17°$ C. Ambient temperatures are generally above this level for most the year.

Increased aeration results in changes in pH and Eh. Changes in pH and Eh would affect iron solubility to a large extent.

Exposure to sunlight could have impact on the growth of bacterial population. This, depending on the growth of either iron oxidising or reducing bacteria, could increase or decrease the iron solubility.

In addition to above, oxidation of pyrites results in increased sulphates and dissolved iron levels. Certain parts of Tomago is known to contain large quantities of pyrites.

There are several processes that could be identified as a result of mining, causing variations in iron levels at Tomago. It is likely, that at any given location either one or more of the processes are in operation. This could be one of the reasons why the increase in iron levels are very site specific.

CONCLUSIONS

Mining of mineral sands at Tomago aquifer has affected groundwater quality by increasing iron levels. Tomago aquifer is also a major source of urban water. If water were to be extracted from the mined area, increased treatment would be required because of higher iron levels.

The mining process changes the equilibrium conditions within the aquifer. These could be due to changes in:

i. Physical,

ii. Chemical, and

iii. Microbiological conditions.

It is likely that at any given location, one or more of these could take place, resulting in very site specific variations in water quality.

REFERENCES

Brassington F.C., (1982) Hydrological problems caused by mining and quarrying. Trans. Inst. Min. Metall. (Section: Appl. Earth Sci.), February, vol 91.
Corbett .R.G. (1977) Effects of coal mining on ground and surface water quality, Monongalia County, West Virginia. The science of the total environment, vol 8., pp 21-38.
Emrich G.H.,Merritt G.L.,(1969) Effects of mine drainage on groundwater. Groundwater 7(3), pp 27 - 32.
Mann A.W., Lintern .M. (1983) Heavy metal dispersion patterns from tailings dumps, Northampton district, Western Australia. Environmental Pollution (Series B), pp 33 - 49.
Mink L.L.,Williams R.E.,Wallace A.T.(1972) Effect of early day mining operations on present day water quality. Groundwater, vol 10, pp 17 - 26.

BEHAVIOR OF AQUIFERS CONCERNING CONTAMINANTS: DIFFERENTIAL PERMEABILITY AND IMPORTANCE OF THE DIFFERENT PURIFICATION PROCESSES

J. Gibert

Hydrobiologie et Ecologie Souterraines, URA, CNRS 367 "Ecologie des Eaux Douces", Université Claude Bernard Lyon 1, 43, Bd du 11 novembre 1918, 69622 Villeurbanne cedex, France

ABSTRACT

Contaminants in transit through groundwater systems cross different types of rocks, each one with different water-bearing characteristics. It is imperative to understand the behavior of different aquifers (porous or karstic) concerning contaminants and the relative functions and mechanisms of major processes in a given system before planning remedial measures. From the surface systems to groundwater four main types of filter can be considered. They intervene most often simultaneously and the interactions are very strong. The activities of the different filters lead us to the concept of permeability, which varies according to the efficiency of the filter, ranging from the most impermeable porous aquifers to karstic aquifers. Among major processes we have chosen to consider three groups: retention, selfpurification and dilution and to discuss their importance and their rate of change. We focus more specifically on the problem in karstic systems. An example concerning nitrate profiles in a French karstic system is given, showing an increase of nitrate concentration in the epikarstic zone due to the impact of agriculture. Very few studies on karstic system contamination are mentioned because the processes controlling contaminant cycling inside the karst are as yet not very well understood.

KEYWORDS

Karstic aquifer; porous aquifer; filter effect; contaminant; nitrate pollution; selfpurification.

INTRODUCTION

In Europe, groundwater is easily available in large quantities and intensive use is made of it (Fried et al. 1989). About 50 % of the freshwater used in France is groundwater. Indeed groundwater has an important role in the drinking water supply. Past experience has shown that not only surface water but also groundwater is subjected to human-related pollution risks. In particular, groundwater quality is threatened by widespread use of substances which involve risks for water, from dumping-grounds, contaminated industrial sites and intensive agricultural activities.

The way in which contaminants are transported and diffused via groundwater depends on the type of aquifer but also on the structure of the soil situated under it. Studies on porous milieus are numerous and recently a symposium has addressed this question concerning "Ecological effects of in situ sediment contaminants" (Hydrobiologia, 149, 1987). Very few studies on karstic systems are mentioned because the processes controlling contaminant cycling inside karsts are not very well understood. However the total karst areas of the world are not negligible, about 5.3 million km^2, i.e. 4 % of the total surface of the continents, excluding Antarctica. Europe is the most karstic continent (13.5 %) compared with North and Central America (3.5 %). The largest karst areas are found in France (about 40 % of the total area) (Balazs, 1977). In these terrains, permeability is high and it is responsible for the short residence times of the water.

Contaminants reach an aquifer by infiltrating downward through an unsaturated (vadose) zone between the surface and the water table, or directly from the surface aquatic system to the groundwater system. A large number of references pertaining to groundwater pollution generally, or to pollution in a particular area or region from multiple causes, have appeared in the literature (Todd and Mc Nulty, 1976). During the last decade much more attention has been focused on the role of the unsaturated zone, on the water/sediment interface, on sampling and monitoring of organic contaminants and on the involvement of minerals in subsurface chemical change (U.N.E.S.C.O., 1983; Brandon, 1986).

The purpose of the paper is to examine the framework of processes and phenomena related to different types of aquifer. The subdivision of aquifers into intergranular (porous) and fissure-flow (karstic) types is a gross over-simplification since most are of an intermediate nature (fig. 1). The studies on groundwater contamination therefore pose many problems which stem from this great range of potential geological conditions, and the almost equally wide range of potential contaminants which can be encountered (Edworthy, 1987). The fate of contaminants is inextricably linked to the behavior of the type of aquifer. We try to evaluate some aspects of behavior of porous and karstic aquifers concerning contaminants, each with its own type of permeability. We focus more specifically on the problem in karstic systems.

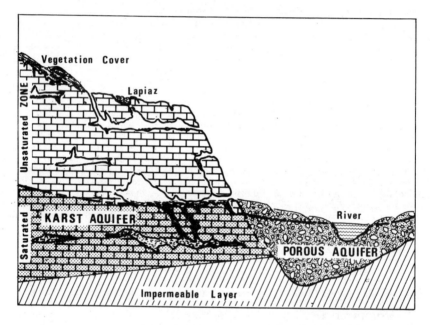

Fig. 1. The two types of aquifer: karstic and porous aquifers with their saturated and unsaturated zones.

DIFFERENTIAL PERMEABILITY

When crossing saturated and unsaturated zones contaminant loads undergo transformations; they may disappear completely, slow down, halt or spread out so widely that they are not easily detected. There are four main types of filter between the surface and groundwater. The first one is of a geological nature, it corresponds to the structure of the matrix, the size of the interstices between sand, gravel and stone conditions through which contaminants flow. This is the mechanical filter. Contaminants can be stopped by sorption in sediments which present a large specific surface, and by photodegradation. Light and other physical factors intervene to form the physical filter. Certain substrates, such as clay or humus exchange ions with those contained in water, and natural and synthetic complexing agents form soluble complexes or colloidal suspensions with organic acids. Under the influence of pH or rH variation during transport, certain solutes precipitate and they are halted. These processes form the chemical filter. Finally contaminants can be degraded by living matter. Microbial activity is very intense in the upper part of the soil, and is increased by the presence of the plant cover. Organisms have a great influence on the fate of contaminants and they form the biological filter.

INTERACTIONS BETWEEN FILTERS

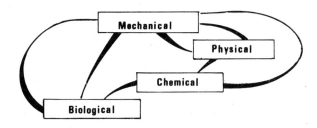

Fig. 2. Interactions of the different filters on the movement and fate of contaminants.

These four filters act most often simultaneously and their interactions are very strong (Fig. 2). For example, when a solute precipitates (chemical filter) it is then stopped by the size of the interstices (mechanical filter) and it can be degraded by organisms (biological filter) on which the physical filter acts. Likewise contaminants may be physically stabilized long enough for chemical and microbiological degradation to occur. The study of the efficiency of the different filters leads to the concept of permeability (Vervier and Gibert, in press). Permeability varies according to the efficiency of the filter, decreasing from the most impermeable porous aquifers to karstic aquifers (Fig. 3). For porous aquifers the four filters - physical, mechanical, chemical and biological - are all important, whereas in karstic aquifers biological and physical filters predominate.

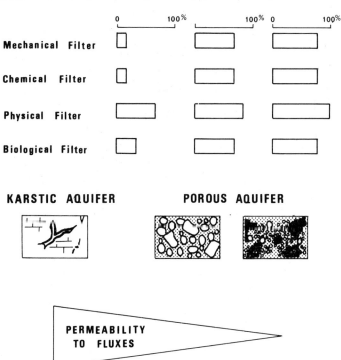

Fig. 3. Proposal for a synthetic succession of the efficiency of the filters from porous aquifer (clogged aquifer) to karstic aquifer and hence of permeability.

IMPORTANCE OF DIFFERENT PURIFICATION PROCESSES

In general major processes controlling contaminant cycling are well known. They are: changes in salt concentration, changes in redox conditions, lowering of pH, complexation of metal ions, changes in turbidity, formation of particulate matter, microbial activity, gas production, benthic invertebrate effect (see Thomas et al. 1987; Förstner, 1987). Among the many processes that occur inside the aquifer we have chosen to consider three mechanisms: retention, selfpurification and dilution and to discuss their importance and their rate of change.

Porous areas

In the porous areas, the unsaturated zone plays an important role in the retention and selfpurification of contaminants (Fig. 4). Considerable physical and chemical changes, most often biologically mediated, reduce the impact of pollution on groundwater quality. Danielopol (1989) suspects that the superficial sediments of the gravel deposits behave as a giant trickling filter and the deeper layers as a slow filter. Periodical fluctuations of the water level could stimulate interstitial bacterial activity in the unsaturated zone.

Retention and release. Large quantities of contaminants are retained in the upper part of the aquifer (soil and surface layers of sediment in the aquatic environment) due to the nature of the substratum and the infiltration speed of contaminants. Fine sediments which in general contain higher contaminant concentration, accumulate in quieter waters such as lagoons, tidal flats, flood plains and deep lakes (Förstner, 1987). The water-sediment and oxic-anoxic interfaces play major roles in the potential flux of pollutants (Salomons et al., 1988). Predictive models of the dynamic network of processes and phenomena in sediment transport have been set up. The clear indications of the model are of cyclical transfer of solids from sink (deposition, settling sediment) to source (erosion, mobile suspended sediment, release) via a variety of relatively ephemeral states where a wide range of inter-related physical, mechanical, chemical and biological processes occur (Parker, 1988).

Selfpurification. These mechanisms concern mainly chemical and biological processes, they degrade contaminants and modify their molecular structure. Many processes control concentrations of selected constituents of solutes in the zone of aeration (such as complexation or oxidation-reduction). Several recent studies have focused on these processes and especially on microbial activity (Beeman and Suflita, 1987) gas production and consumption, processes effected by invertebrates (see Reynoldson, 1987; Fukuhara and Sakamoto, 1988) such as bioaccumulation through ingestion and biomagnification (real biological amplification of the chemical pollution due to transfer and contaminant concentration along trophic webs), and bioturbation. In alluvial groundwater, the large number of communities situated in the upper part of the aquifer (down to a depth of 3 m) form the biological layer (Dole and Mathieu, 1984) and can play a major role in purification. In all the studies on invertebrates, the stygobiont fauna has been neglected. However, for instance, Sinton (1984) believes that the presence of stygobiont fauna, mainly Amphipods and Isopods, in the wells of a sewage treatment plant is linked to the abundance of organic contaminants. He calculates that fauna can recycle 20 % of the calorific value of the output of the sewage plant. It is clear that further studies are required in order to better establish trophic inter-relationships in polluted and unpolluted aquifer ecosystems.

Dilution. This is one of the few clean-up mechanisms which takes place in the saturated zone. It is essentially dispersion which diminishes concentration during transfer. Hydrological models have been set up and they emphasize the importance of a thorough knowledge of the links existing between the aquifer and the surface network (Pinder, 1984; Jennings, 1987).

Karstic systems

In the karstic aquifer, heterogeneity is observed whatever the scale seak. Flow properties of the aquifer are very different inside or outside the drains (directional hydraulic conductivity) (Mangin, 1986). As far as contaminants are concerned, filtration by surface karst is only rare and weak. However in a wooded karstic catchment area, contaminant input-output balances reveal that 87.5 to 99.7 % of the atmospheric micropollutants released from the snow cover are retained in the karst system (Simmleit and Herrman, 1987). Contaminants can short-circuit the unsaturated zone entirely because of the structure of the karst (wide crevices, lack of filtering structures) and because the soils form thin layers. Most important processes take place in the saturated zone (Fig. 4).

Dilution. The higher the discharge rate of the groundwater (or the speed, or the transmissibility), the greater the dilution. Contaminants are eliminated more easily if the water is renewed frequently. This is true for non-intertial karsts. Frequent floods can lead to the mixing of different water layers and improve dilution. For soluble substances, the dilution phenomenon intervenes favorably in large watersheds.

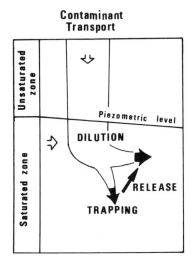 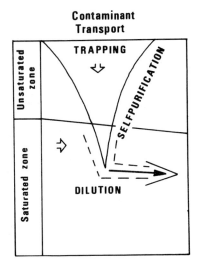

Fig. 4. Importance of the three processes: selfpurification, dilution and trapping (retention) in karstic and porous aquifers, concerning the satured and unsaturated zones.

Retention and Release. The trapping processes in the karst pose many problems because of their temporary nature which implies concentrated release in proportions exceeding the original input. Pollutants are retained by sedimentation (adsorption in the clay), by gravity (at the bottom of a well or in a siphon) or by flotation (oil, for example) (Maire, 1979). They are released rapidly and totally or in a deferred concentrated way (especially after high water).

Selfpurification. The best conditions for natural selfpurification are not found in karstic systems. There is not enough time for the karstic groundwater to be throughly oxygenated before it reaches the saturated zone (in a few hours or a few days) where selfpurification can only take place anaerobically. The low density of microorganisms reduces the rate of mineralisation of organic substances and the almost total lack of algae prevents the mobilization of part of these pollutants (nitrates) when they are overabundant. The cave-dwelling populations may either increase (Holsinger, 1966; Turquin, 1980; Sinton, 1984) or diminish (Brantsetter, 1975; Adamek and Rauser, 1977) when these pollutants reach them. Heavy metals may accumulate in large quantities in cave-dwelling animals, e.g. in crayfish in Tennessee (U.S.A.) (Dickson et al., 1979). Evaluation of heavy metal contents revealed that epigean crustacean communities are more sensitive to toxicants (cadmium and zinc) than cave-dwellers but there is no difference in reaction to free and combined chlorine (Bosnak and Morgan, 1981). Biotransformation and biodegradation by microbial or macrofaunal action are not well known inside the karst and need more attention (Turquin, 1980; Moreau, 1982).

The implications of water quality stratification in karstic aquifers can be illustrated by the range of nitrate concentrations found in the Dorvan Cleyzieu massif (Fig. 5). It is situated on the southwestern outer range of the French Southern Jura mountains and forms a stepped karstic unit. It is characterized in the upper zone by an epikarstic aquifer with a main sub-surface channel, corresponding to the stream and to the exsurgence of the Cormoran cave (steady flow, vadose regime), and in the lower zone by a main channel that emerges at the Pissoir Spring in the Albarine alluvial plain (saturated regime, temporary flow). The relationships between the different zones have been demonstrated by a multitracer experiment (Gibert, 1986). The geological structure of this aquifer is formed by the middle Jurassic limestone (Bathonian and Bajocian).

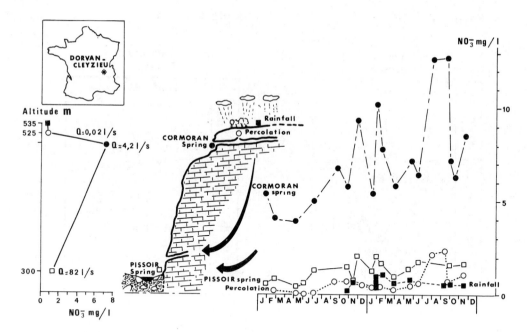

Fig. 5 Schematic section of nitrate distribution in relation to land use and flow system in a French Jurassian karst. Q = mean discharge.

A vertical geochemical gradient was shown to exist throughout the massif; the mineralisation of the water of the saturated zone is lower in the top layers. The solute concentration (Ca^{++}, Mg^{++}, HCO_3^- and SO_4^{--}) increases gradually, and so the water of the lower aquifer contains more ions (between 300 and 400 µS) than that of the upper aquifer (between 100 and 285 µS for percolations and between 170 and 370 µS for the Cormoran Stream). Seasonal variations were clearly evident in the water of the epikarstic aquifer and this phenomenon is limited to the proximity of the ground surface. The impact of agriculture on the quality of the groundwater is shown in Cormoran Spring for Cl^- (up to 8 mg.l-1) and for NO_3^- (up to 14.7 mg.l-1). So far, this contamination has remained weak and confined to a small area for the non-saturated zone (epikarstic area) (Fig. 5). Percolation sites localized under the forest do not present high concentrations of these organic contaminants (0.7 mg.l^{-1}). The same is true for the saturated zone (1.3 mg.l^{-1}). Studies carried out over several successive years do not show important changes in the profiles (Gibert, 1986). Differences in nitrate concentrations inside this karst, could be explained by dilution, but it is possible that selfpurification (denitrification) takes place inside the aquifer, leading to a restoration of the water quality at the base level. Other studies in small, rural karst catchments have shown that nitrates are concentrated not only in the epikarst, but can also be found in the deepest layers of the karst (Simmeleit and Hempfling, 1986; Simmeleit, 1988). Deforestation and intensive farming are likely in the more-and-less long term to contaminate the deepest storage areas of karsts.

CONCLUSION

The recent French-German symposia on water and soils (Strasbourg, March 1988), and on the contamination of groundwater by nitrates (Stuttgart, October 1988) have shown the need to take into account simultaneously the incidence of pollution on groundwater quality and its impact on the different milieus. It is not possible to study the problem of the fate of contaminants properly without taking into account the structure of aquifers.

Groundwater is very unequally exposed to pollution risks. The priority given to protection and to groundwater quality in porous environments should not let us forget that karstic groundwater will be used more and more in the future. The karstic environment has been studied for only a few years and its functioning is not completely understood, so more studies, using new technology and methods, should be undertaken.

ACKNOWLEDGEMENTS

It is a pleasure to thank Dominique Martin and Roger Laurent for their competent assistance with field work and for technical assistance in the preparation of the manuscript and Nathalie Lyvet for kindly typing the manuscript. Particular thanks are due to Glynn Thoiron for editing this text.

REFERENCES

Adamek, Z. and Rauser, J. (1977). Contribution to the question of the water quality of the Moravian karst on example of the mayflies (Ephemeroptera) and stoneflies (Plecoptera) fauna. Speleol. Vestnik (Brno), 3, 7-23.

Balazs, D. (1977). The geographical distribution of karst areas. Proc. of the 7th Int. Speleological Cong., Sheffield, England, 13-15.

Beeman, R.E. and Suflita, J.M. (1987). Microbial ecology of a shallow unconfined ground water aquifer polluted by municipal landfill leachate. Microb. Ecol., 14, 39-54.

Bosnak, A.D. and Morgan, E.L. (1981). Acute toxicity of cadmium, zinc and total residual chlorine to epigean and hypogean isopods (asellidae). N.S.S. Bulletin, 43, 12-18.

Brandon, T.W. (1986). Water practice manual, 5. Groundwater: occurrence, development and protection. Institute of Water Engineers and Scientists, 1-615.

Brantsetter, J. (1975). A reconnaissance investigation of pollution in a limestone terrane, Horse Cave, Kentucky. The Kentucky Caver (Lexington), 9,(3), 39-41.

Danielopol, D.L. (1989). Groundwater fauna associated with riverine aquifers. J.N. Am. Benthol. Soc., 8, (1), 18-35.

Dickson, G.W., Briese, L. and Giesy, J.P. Jr. (1979). Tissue metal concentrations in two crayfish cohabiting a Tennessee cave stream. Oecologia (Berl.), 44, 8-12.

Dole, M.J. and Mathieu, J. (1984). Etude de la "pellicule biologique" dans les milieux interstitiels de l'Est lyonnais. Verh. Internat. Verein. Limnol., 22, 1745-1750.

Edworthy, K.J. (1987). Groundwater contamination. A review of current knowledge and research activity. Stygologia, 3,(4), 279-295.

Fried, J.J. and Zampetti, M. (1989). Les ressources en eaux souterraines de la Communauté Européenne: leur inventaire, leur vulnérabilité, leur qualité. Hydrogéologie, 2, 71-74.

Förstner, U. (1987). Sediment-associated contaminants - an overview of scientific bases for developing remedial options. Hydrobiologia, 149, 221-246.

Fukuhara, H. and Sakamoto, M. (1988). Ecological significance of bioturbation of zoobenthos community in nitrogen release from bottom sediments in a shallow eutrophic lake. Arch. Hydrobiol., 113,(3), 425-445.

Gibert, J. (1986). Ecologie d'un système karstique jurassien. Hydrogéologie, Dérive animale, Transits de matières, Dynamique de la population de Niphargus (Crustacé, Amphipode). Mémoires de Biospéologie, 13,(40), 1-379.

Holsinger, J.R. (1966). A preliminary study on the effects of organic pollution of Banners Corner Cave, Virginia. Int. J. Speleol., 2, 75-98.

Jennings, A. (1987). Critical chemical reaction rates for multicomponent groundwater contamination models. Water Resources Research, 23,(9), 1775-1784.

Maire, R. (1979). Comportement du karst vis-à-vis des substances polluantes. Annls Soc. Géol. Belgique, 102, 101-108.

Mangin, A. (1986). Réflexion sur l'approche et la modélisation des aquifères karstiques. Journées sur le karst en Euskadi 86. Donostia, San-Sebastian, 11-31.

Moreau, R. (1982). La pollution des eaux souterraines: hygiène et épidémiologie. Journée d'étude sur la protection des eaux karstiques. Société Nationale des Distributions d'Eau et Commission Nationale de Protection des Sites Spéléologiques, 4-32.

Morel-Seytoux, H.J. (1988). Soil-aquifer-stream interactions- a reductionist attempt toward physical-stochastic integration. J. of Hydrology, 102, 355-379.

Parker, W.R. (1988). On the role of fine sediment behaviour in pollutant transfer modelling. Wat. Sci. Tech., 20,(6/7), 175-182.

Pinder, G.F. (1984). Groundwater contamination transport modeling. Environmental Science and Technology 18,(4), 108-114.

Reynoldson, T.B. (1987). Interactions between sediment contaminants and benthic organisms. Hydrobiologia, 149, 53-66.

Salomons, W., de Rooij, N.M., Kerdijk, H. and Bril, J. (1987). Sediments as a source for contaminants, Hydrobiologia, 149, 13-30.

Simmleit, N. (1988). Variations of inorganic and organic composition of bulk precipitation, percolation water and groundwater in small, rural karst catchments. Catena, 15, 195-204.

Simmleit, N. and Hempfling, R. (1986). Stickstoffmineralisation und Nitratbilanz in Karsteinzugsgebieten der Nördlichen Frankenalb. Wasser + Boden, 38, 608-613.

Simmleit, N. and Herrman, R. (1987). The behavior of hydrophobic, organic micropolluants in different karst water systems. 1. Transport of micropollutants and contaminant balances during the melting of snow. Water, Air and Soil Pollution, 34, 79-95

Sinton, L.W. (1984). The macroinvertebrates in a sewage-polluted aquifer. Hydrobiologia, 119, 161-169.

Thomas, R., Evans, R., Hamilton, A., Munawar, M., Reynoldson, T. and Sadar H. (1987). Ecological effects of in situ sediment contaminants. Hydrobiologia, 149, 259-272.

Todd, D.J. and Mc Nulty, D.E.O. (1976). A review of the significant literature on polluted groundwater. Water Information Center Publication, Huntington, New York

Turquin, M.J. (1980). La pollution des eaux souterraines: incidence sur 'es biocénoses aquatiques. Actes 1° Coll. Nat. Protection Eaux Souterraines Besançon, Les Cahiers de la C.P.E.P.E.S.C., 2, 341-347.

U.N.E.S.C.O., (1983). Aquifer contamination and protection. Studies and reports in hydrology, Publication n°30, 1-440.

Vervier, P. and Gibert, J. (submitted). Permeability to nutrient fluxes in freshwater/groundwater ecotones. Hydrobiologia.

SUBJECT INDEX

Alachlor, soil mobility 87-94
Alkylbenzenes
    adsorption, and head space analysis 7-14
    partitioning 7-14
Aquifers
    contamination with hydrocarbons 27-36
    and groundwater contamination 101-108
    and mineral sand extraction 95-100
    types, and groundwater contamination 101-108
Atrazine, soil mobility 87-94
Aviation gasoline
    fate and transport, groundwater 37-44
    immiscible, groundwater pollution 37-44
    separate phase 37-44

Bentonite, permeability 23-26
Benzene, biodegradation 53-62
Biodegradation
    gasoline in groundwater 53-62
    see also named contaminants
Biorestoration
    gasoline contaminated groundwater 53-68
    gasoline contaminated soil 63-68
    oil contaminated soil and groundwater 63-68
Biotreatment, enhanced *in situ*, oil contaminated soil and groundwater 63-68
BTEX, biodegradation, subsurface materials contaminated with gasoline 53-62

Complexation-adsorption model, mobility of hydrophobic compounds, in groundwater 15-22
Contaminants
    organic, sorption 1-6
    transport in groundwater 101-108
    see also named contaminants

Ethylbenzene, biodegradation 53-62

Gasoline
    groundwater contamination 53-62
    soil contamination, *in situ*, biotreatment 63-68
    storage tank leakage, and subsurface environment contamination 53-62
Groundwater
    aviation gasoline pollution 37-44
    contamination 101-108
    and sorption of organic contaminants 1-6
    gasoline contamination 53-62
    herbicide pollution 87-94
    hydrocarbon contamination 27-36
    biotreatment 63-68
    hydrophobic compounds, mobility 15-22
    iron levels 95-100
    pesticide contamination 87-96
    and phenoloxidases, in chemical pollution treatment 69-77
    pollution, organochlorines 79-86

Headspace analysis, and alkylbenzene adsorption to mineral oxides 7-14
Herbicides
    groundwater contamination 87-94
    transport in soil 87-96
Hydrocarbons
    mobility, in aquifers 27-36
    and soil contamination, biotreatment 63-68
Hydrophobic compounds, mobility, and influence of dissolved organic matter 15-22

Ilmenite, aquifer content 95-100
Iron
    groundwater levels 95-100
    mineral sand levels 95-100
    removal, from water 95-100

Karstic aquifer, and groundwater contamination 101-108

Subject index

Metolachlor, soil mobility 87-94
Mineral oxides, and alkylbenzene adsorption 7-14
Mineral sands, and iron levels 95-100
Mining, mineral sand 95-100

Natural sorbents, and sorption of organic contaminants 1-6
Nitrate, groundwater contamination 101-108
Non point pollution 87-96

Oil, soil contamination, *in situ* biotreatment 63-68
Organic contaminants
    partitioning 1-6
    in soil and groundwater 1-6
    sorption 1-6
Organic matter, in groundwater, and mobility of hydrophobic compounds in groundwater 15-22
Organic pollutants, adsorption 23-26
Organochlorines
    groundwater contamination 79-86
    monitoring 79-86

Pesticides, groundwater contamination 87-96
Petroleum
    contamination soil
    experimental model 45-52
    fate and transport 45-52
Petroleum products, and aquifer contamination 27-36
Phenoloxidases
    activity inhibition by soil related adsorbents 69-77
    in chemical pollution treatment 69-77
Porous aquifer, and groundwater contamination 101-108

Quaternary ammonium cations, and bentonite permeability in a stabilization pond liner 23-26

Rutile, aquifer content 95-100

Soil
    adsorbents, and inhibition of phenoloxidase activity 69-72
    gas analysis 79-86
    hydrocarbon contamination, biotreatment 63-68
    petroleum contamination 45-52
    and phenoloxidases, in chemical pollution treatment 69-77
    types, and petroleum contamination 45-52
Sorption, alkylbenzenes, to mineral oxides 7-14
Stabilization pond liner, and organophilic bentonites 23-26
Surfactants, and gas-oil recovery enhancement 27-36

Tetrachloroethylene, groundwater pollution 79-86
Toluene, biodegradation 53-62
Tomago sandbeds, Australia 95-100
Trichloroethylene, groundwater pollution 79-86

Vapour-phase sorption, organic contaminants 1-6
Vinasse
    organic components, adsorption, bentonite 23-26
    treatment, stabilization pond liner 23-26

Xenobiotics
    microbial degradation 69-77
    persistence in soil and groundwater 69-77
Xylene, biodegradation 53-62

Zircon, aquifer content 95-100